Current Topics in Microbiology and Immunology

82

Edited by

W. Arber, Basle · W. Henle, Philadelphia · P.H. Hofschneider,
Martinsried · J.H. Humphrey, London · J. Klein, Tübingen · P. Koldovský,
Düsseldorf · H. Koprowski, Philadelphia · O. Maaløe, Copenhagen ·
F. Melchers, Basle · R. Rott, Gießen · H.G. Schweiger, Ladenburg/Heidelberg ·
L. Syruček, Prague · P.K. Vogt, Los Angeles

With 13 Figures

Springer-Verlag
Berlin Heidelberg New York 1978

ISBN-13: 978-3-642-46390-7   e-ISBN-13: 978-3-642-46388-4
DOI: 10.1007/978-3-642-46388-4

2121/3321-543210

Table of Contents

Indexed in Current Contents

Anti-Actin Antibodies

ASTRID FAGRAEUS and RENÉE NORBERG [1]

I. Introduction

Although antibodies against actin were claimed to have been experimentally produced by *Kesztyüs* et al. (1949) and subsequently by several others (*Marshall* et al., 1959; *Tunik* and *Holtzer*, 1961; *Pepe*, 1966; *Hirabayashi* and *Hayashi*, 1972), the specificity of the antisera produced was nevertheless questioned (*Bray*, 1974). Actin, which is present in almost every cell of all animals, has even been considered nonantigenic because of its strongly conserved structure with very small differences between species. Thus, all animals should be tolerant, and this would explain the many unsuccessful attempts to produce antisera against actin.

In the last few years, however, the production of actin antibodies in experimental animals has been convincingly demonstrated (*Trenchev* et al., 1974; *Lazarides* and *Weber*, 1974; *Lazarides*, 1975; *Trenchev* and *Holborow*, 1976). Moreover, antibodies against actin have been shown to occur in many human diseases.

Interaction between actin and other contractile proteins provides the molecular basis of motility in muscles but also in numerous other biologic systems. Actin is present within the muscles in an easily recognized, well-organized form, whereas in nonmuscle cells it can be demonstrated in at least two states, as bundles of filaments and as poorly defined, diffusely distributed meshwork. It is obvious that the mesh can convert into filaments and vice versa by, e.g., contact-mediated signals emanating from the cell's periphery.

---

[1] Department of Immunology, The National Bacteriological Laboratory, 10521 Stockholm, Sweden

Although the cell's locomotive system might be influenced by preparative methods and by fixation procedures necessary for immunofluorescence experiments (IFL), anti-actin antibodies are valuable tools in studying the contractile machinery under various conditions.

This review will deal with anti-actin antibodies experimentally produced in animals or spontaneously occurring in man. The production and characteristics of anti-actin antibodies of various origins will be compared and the applications hitherto most used will be related. Moreover, the use of human anti-actin antibodies as a diagnostic tool will be considered.

II. Experimentally Produced Anti-Actin Antibodies

The first attempt to produce anti-actin serum experimentally was made by *Kesztyüs* et al. (1949). They produced antibodies in two out of three rabbits by immunizing them with rabbit skeletal actin. The antisera precipitated the antigen used for immunization but not a myosin preparation.

Later on, several others (*Marshall* et al., 1959; *Tunik* and *Holtzer*, 1961; *Pepe*, 1966; *Hirabayashi* and *Hayashi*, 1972; *Wilson* and *Finck*, 1971) prepared antisera to skeletal muscle actin, but the monospecificity of these sera has been questioned. Since 1974, however, several anti-actin sera have been produced and characterized according to specificity and reactivity with various kinds of cells (*Trenchev* et al., 1974; *Lazarides* and *Weber*, 1974; *Lazarides*, 1975; *Trenchev* and *Holborow*, 1976; *Owaribe* and *Hatano*, 1975; *Fagraeus* et al., 1977).

A. Immunizing Antigen; Source and Preparation

Actins of various species and tissues seem to be very similar in several of their physiologic, physical and chemical properties. Amino sequence analyses have shown very small differences, e.g., between skeletal muscle actin of rabbit and actin from fish (*Elzinga* et al., 1973; *Collins* and *Elzinga*, 1975).

Within the same species, actins from cardiac and skeletal muscle are essentially indistinguishable and the sequence differences observed in these actins do not seem to affect their functional characteristics. But, it is nevertheless clear that there are several actins within a species and that muscle and cytoplasmic actins are slightly different in amino acid sequences (*Elzinga* et al., 1976). Moreover, a large fraction of actin present in, e.g., brain and fibroblasts (*Bray* and *Thomas*, 1976), as well as in platelets (*Gallagher* et al., 1976), differ functionally from muscle actin. There is, on the other hand, no proven antigenic difference between actins of different species and tissues. Antisera against skeletal muscle actin react with actin-containing structures of nonmuscle cells and vice versa (*Lazarides*, 1975; *Trenchev* and *Holborow*, 1976; *Fagraeus, Biberfeld*, and *Norberg*, 1977). *Owaribe* and *Hatano* (1975), however, have reported an exception as their rabbit antiserum against actin from Plasmodium myxomycete did not react with actin from rabbit skeletal muscle.

The immunizing actins have been isolated according to standard procedures (*Fagraeus* et al., 1977) (Fig. 1) and further purified by gel filtration (*Trenchev*

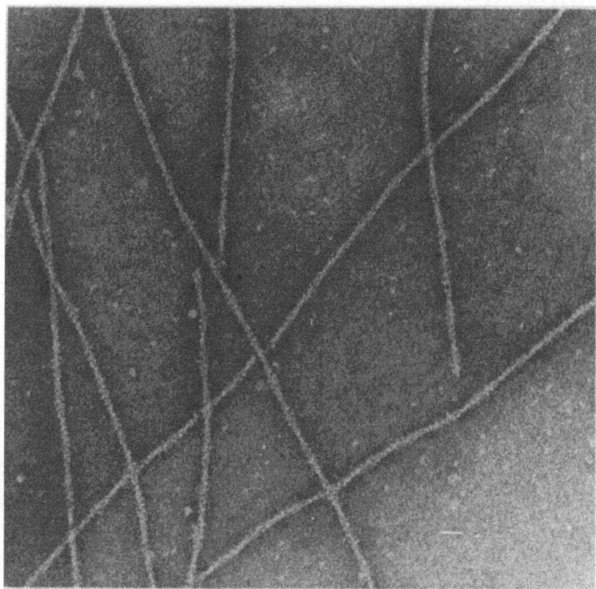

Fig. 1. Electron micrograph showing F-actin prepared from rabbit skeletal muscle. Negative staining with 2% sodiumtungstosilicate. × 150 000

et al., 1974; *Owaribe* and *Hatano,* 1975; *Trenchev* and *Holborow,* 1976). A very elegant method was introduced when *Lazarides* and *Weber* (1974) isolated actin from mouse fibroblasts and purified it further through sodium dodecyl sulfate (SDS) slab gel electrophoresis. The proteins comigrating with actin on SDS slab gels were recovered from the gel by elution and then used as immunogen.

Actin is considered to be a poor antigen, which has been attributed to its widespread distribution in nature. However, actin denatured by SDS or by storage for at least a week at $+4°$ C is considered more immunogenic than a native preparation. It is also the experience of many researchers that only a restricted number of animals produced antibodies after immunization. According to our research (*Fagraeus* et al., 1977) one out of ten rabbits immunized with rabbit skeletal muscle actin produced antibodies in satisfactory amounts. *Trenchev* and *Holborow* (1976) found precipitating antibodies in 14 out of 36 rabbits injected with human smooth muscle actin. They did not mention if some animals produced nonprecipitating antibodies (see below).

The amount of antigen used has varied from 0.5–10 mg per injection, given in *Freund's* (complete or incomplete) adjuvant at different intervals. When testing the immunogenicity of actin it is necessary to take pre-immunization serum samples since the incidence of spontaneously occurring antibodies reacting with muscles and actin-containing structures in nonmuscle cells is high in rabbits. The reason for this antibody production is unknown. It seems to vary with the rabbits' origin and might be related to various virus infections affecting the animals (see below).

III. Human Anti-Actin Antibodies

Human antibodies (SMA) reacting with smooth muscle antigens were first described by *Johnson* et al. (1965) in patients with chronic active hepatitis (CAH). SMA-positive sera from CAH-patients were also shown to react with renal glomeruli (*Whittingham* et al., 1966), liver cells (*Farrow* et al., 1971), thyroid cells (*Biberfeld* et al., 1974; *Sutton* et al., 1974), the brush border of renal tubular and intestinal epithelial cells (*Gabbiani* et al., 1973; *Fagraeus, Lidman,* and *Norberg,* 1975), microfilaments in fibroblasts grown on glass (*Gabbiani* et al., 1973), lymphoid cells (*Fagraeus* et al., 1973), and platelets *(Gabbiani* et al., 1972; *Norberg* et al., 1975). (Fig. 2)

A. Actin Specificity of SMA

The specificity of SMA characterized by the broad reactivity described above was unknown until *Gabbiani* et al. (1973) were able to abolish the staining of SMA-positive sera by absorption with thrombosthenin A, the actin moiety of platelets.

The actin specificity of SMA-positive sera, however, was not generally accepted until 1976 when the results were confirmed by adsorbing such sera with rabbit skeletal muscle actin (*Lidman* et al., 1976; *Chaponnier* et al., 1976), actin prepared from human uterus or ox stomach (*Botazzo* et al., 1976) and actin from rabbit skeletal muscle or gizzard muscle (*Andersen* et al., 1976). In all experiments absorptions with other contractile proteins prepared from smooth or striated muscle were unsuccessful.

Broad-reacting SMA stain the actin-rich I-bands of isolated skeletal muscle myofibrils *(Andersen* et al., 1976; *Chaponnier* et al., 1977). Analysis through SDS slab gel electrophoresis of precipitates formed between actin and SMA demonstrated only bands corresponding to IgG and actin (*Lidman* et al., 1976; *Utter* et al., 1977). By electronmicroscopy, *Utter* et al. (1977) studied the complexes formed by F-actin and broad-reacting SMA. The immune complexes consisted of parallel arrays of actin filaments cross-linked by antibodies. Thus, the actin specificity of SMA-positive sera from patients with CAH seems to be well-documented.

The actin specificity of other SMA-positive sera, however, has so far been confirmed only in a restricted number of sera, mostly from patients with liver diseases and with EB virus infections (*Lidman* et al., 1976; *Botazzo* et al., 1976; *Lamelin* et al., 1977), although broad-reacting SMA are found under many other conditions (*Whitehouse* and *Holborow,* 1971; *Holborow* et al., 1973; *Andersen,* 1975; *Wasserman* et al., 1975; *Biberfeld* and *Sterner,* 1976; *Lidman,* 1976; *Lamelin* et al., 1977). In this context it should be mentioned that not all human sera reacting by IFL with smooth muscle exhibit actin specificity. Several other specificities are known (*Lidman* et al., 1976; *Andersen* et al., 1976; *Botazzo* et al., 1976; *Fairfax* and *Groeschel-Stewart,* 1977). Most of the latter sera, however, do not give the broad cell-staining pattern shown by anti-actin sera. Thus, it is necessary to perform thorough analyses including absorption experiments before establishing the actin specificity of human SMA.

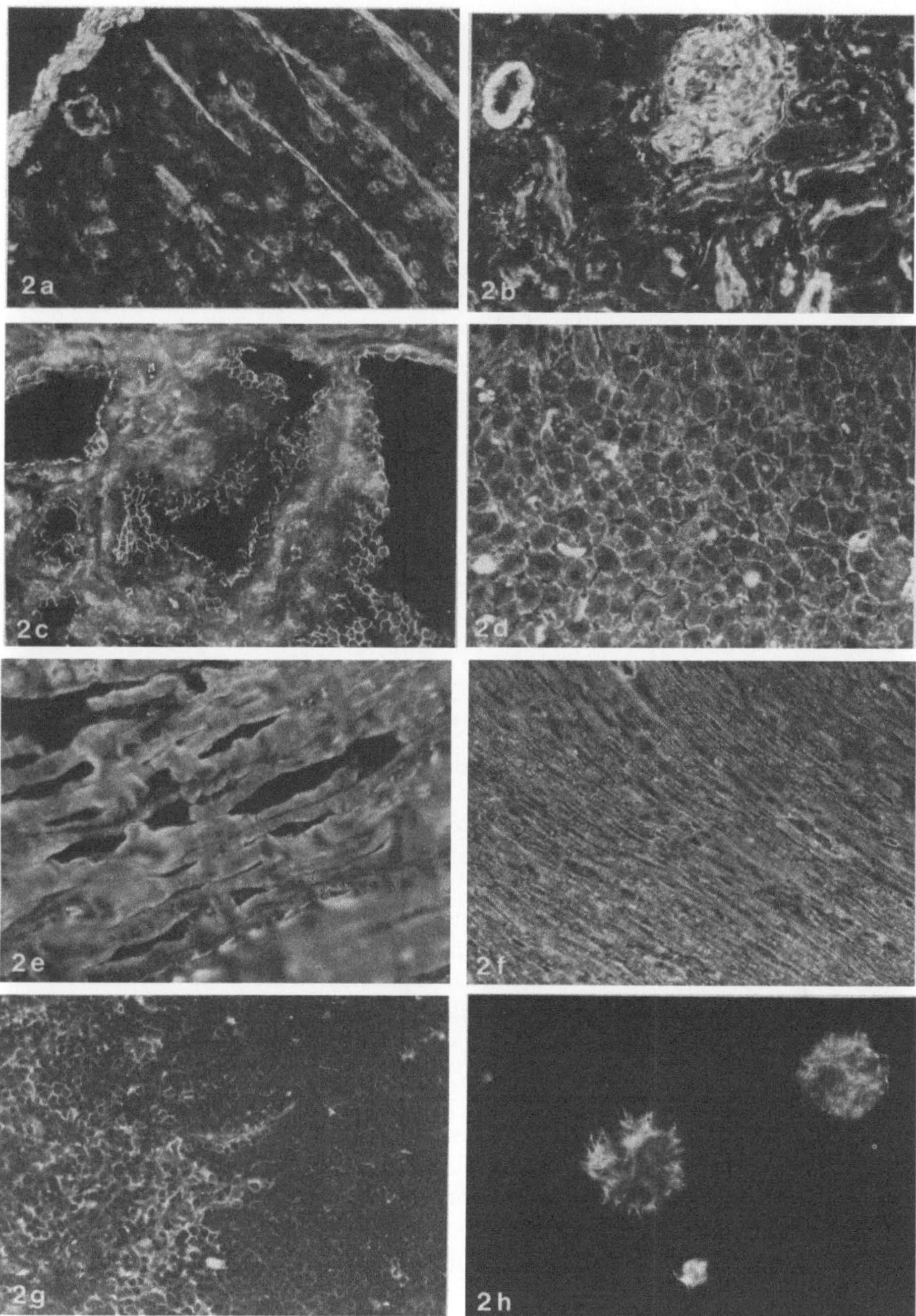

Fig. 2a–h. Indirect immunofluorescence experiments showing staining produced by human anti-actin SMA-positive serum. (a) Positive reaction in muscularis mucosae and the smooth muscle fibres between the mucosal cells in sections of rat stomach. ×325. (b) Staining of walls of blood vessels, glomerulus, the brush borders of proximal renal tubules and basal parts of tubular cells of rat kidney. ×325. (c) Fluorescence in the membrane region of the epithelial cells in sections of human thyrotoxic thyroid. ×325. (d) Pericellular, "polygonal" pattern of sections of rat liver cells. ×325. (e), (f) Striated pattern on monkey skeletal and rat cardiac muscle. ×325. (g) Staining of medullary cells of sections of mouse thymus. ×325. (h) Fluorescent surface microvilli of smeared lymphoblastoid cells. Note also a free bunch of "hairs". ×1000

B. Incidence and Diagnostic Significance
of Human Anti-Actin Antibodies

In healthy human controls the incidence of SMA, demonstrated by immunofluorescence techniques, has been 3%–18% (*Wasserman* et al., 1975; *Shu* et al., 1975; *Lidman* et al., 1976; *Biberfeld* and *Sterner,* 1976). The titers are low (< 25). We found, however, in immunodiffusion experiments according to Ouchterlony that 30 out of 100 undiluted blood donor sera precipitated actin prepared from rabbit skeletal muscle (unpublished results).

High titers (≥ 80) of SMA with actin specificity (*Gabbiani* et al., 1973; *Lidman* et al., 1976; *Andersen* et al., 1976; *Botazzo* et al., 1976) are essentially found in patients with CAH where the incidence in Caucasian patients with a definite diagnosis is 70% or more (*Whittingham* et al., 1966; *Galbraith* et al., 1974). Some studies (*Whittingham* et al., 1966; *Olsson* and *Hultén,* 1975; *Lidman* et al., 1976) have shown that in CAH the anti-actin titers varied at least to some extent according to the disease activity. Thus, clinical impairment was mostly accompanied by increased anti-actin titers, whereas after successful therapy the titers mostly decreased or even disappeared. On the other hand, *Murray-Lyon* et al. (1973), in their controlled trial of treatment in adult patients with CAH, found no apparent relationship between changes in titers of SMA and treatment or alterations in liver function tests.

Broad-reacting SMA are occasionally found, usually at low titers (< 80) in a variety of chronic diseases with or without liver involvment (*Lidman,* 1976). The actin specificity of these sera, however, has hitherto been documented only in a minority of cases. With actin prepared from rabbit skeletal muscle *Lidman* (1976) absorbed out all SMA activity of 20 sera randomly selected out of 49 SMA-positive sera from patients with nonliver diseases. Anti-actin antibodies found in CAH and other chronic diseases are mainly of IgG class. Their diagnostic importance is obvious, especially in liver diseases where the finding of high titers might indicate a chronic active hepatitis.

Broad-reacting SMA, mostly of IgM class have been reported in various acute viral diseases. *Farrow* et al. (1970) found SMA in 87% of patients with acute viral hepatitis. The antibodies occurred both in HB_sAg-positive and -negative cases. The incidence of SMA was highest during the first month after onset of symptoms and disappeared gradually thereafter. *Andersen* et al. (1976) found that the specificity of IgM–SMA in sera from patients with acute hepatitis varied. Some sera were completely absorbed out by F-actin, others were not.

In cases of *Paul–Bunnell* positive infectious mononucleosis, *Holborow* et al. (1973) reported the finding of SMA in 81% of patients tested within one month of the onset of symptoms. Similar figures have later on been published by *Sutton* et al. (1974) and *Andersen* (1975). SMA have also been demonstrated in measles and mumps (*Haire,* 1972; *Biberfeld,* personal communication), cytomegalovirus infections (*Ajdukiewicz* et al., 1972; *Andersen* and *Andersen,* 1975), and wart virus infections (*McMillan* and *Haire,* 1975). The only nonviral, acute infection where an increased incidence of SMA has been described is *Mycoplasma pneumoniae* infection (*Biberfeld* and *Sterner,* 1976).

C. Etiology of Spontaneously Occurring Anti-Actin Antibodies

Anti-actin antibodies may develop in most individuals in certain viral infections. The presence of virus is in some way of crucial importance for the anti-actin production, since these antibodies are only exceptionally demonstrable in connection with cell damage caused by other agents. *Li* et al. (1977) were able to produce anti-actin antibodies—and antibodies reacting with heavy meromyosin—in rats by ligation of a liver lobe or by cryosurgical damage to a liver lobe provided the damaged tissue was not removed. Other methods causing cell necrosis did not cause production of SMA. The actin-virus complex may in some way act like a hapten-carrier complex. The recent finding of actin, derived from the host cell within RNA tumor viruses (*Lamb* et al., 1976; *Wang* et al., 1976) and paramyxoviruses (*Wang* et al., 1976; *Tyrrell* and *Norrby,* 1977), might be of relevance in this context. The production of anti-actin antibodies in acute viral infections might well be elicited by actin-virus complexes. Additional evidence for the importance of the hapten-carrier theory is the appearance of SMA after drug therapy (*Reynolds* et al., 1971). *Lidman* (1976) showed the actin specificity of these SMA in two patients with hepatitis after oxyphenisatin therapy. After viral infection and also after withdrawal of oxyphenisatin the antibodies gradually disappear. Thus, the occurrence of anti-actin antibodies at low titer in sera from normal human subjects and normal rabbits may be a consequence of earlier viral infections.

In patients with CAH of unknown etiology there is mostly a continuous production of large amounts of IgG anti-actin antibodies. These patients, however, might have a genetic predisposition for an abnormal immune response related to the observed increased frequency of HL-A 1 and 8 in patients with CAH (*Mackay* and *Morris,* 1972).

The anti-actin antibodies are not considered to be directly involved in liver cell destruction as they do not react with the surface of normal living cells (*Farrow* et al., 1971; *Gabbiani* et al., 1973; *Fagraeus* et al., 1975; *Andersen* et al., 1976) and as there is no evidence that SMA are cytotoxic to liver cells (*Paronetto* et al., 1973).

IV. The Use of Anti-Actin Antibodies as a Diagnostic Tool

The interpretation of staining patterns obtained by direct or indirect immunofluorescence is closely connected with the properties of the reactants used. This is valid for all systems but perhaps especially for experiments involving cellular contractile proteins and anti-actin antibodies. It might be discussed whether the experimental conditions permit a preservation of the cellular contractile proteins in their native state. Although important biochemical characteristics are shared by actins from muscle and nonmuscle cells, it is evident that their properties within the cells may differ and that therefore in vitro treatment, e.g., fixation, may give different results depending on the type of cell investigated.

A. Preparation of Test Material

Ordinary cryostat sections of various tissues have mostly been used unfixed, or fixed in dry acetone. Cells grown on glass were fixed in dry acetone at $-20°$ for 20 min (*Norberg* et al., 1975) or used fixed for 30 min in 3.5% formaldehyde in PBS followed by various time periods in acetone (*Lazarides*, 1976). When making preparations of suspended cells, e.g., lymphoid cells and platelets we have shown (*Fagraeus* et al., 1975) that a prerequisite for obtaining a positive staining reaction with anti-actin serum was to deprive the suspending medium of Ca^{2+}, e.g., by using a chelating agent. The reason for this is unknown but might be connected with the organization of cellular actin in the presence or absence of Ca^{2+}. A suitable suspending medium was 0.034 M sodium citrate. Smears on glass slides were fixed for 20 min in dry acetone at $-20°$ C.

Within the muscle cell actin constitutes the thin filaments – the I-bands – which are sharply stained by anti-actin antibodies. Within the nonmuscle cells the organization of actin seems to vary both within a specific cell and between cells of various kind. Several investigations indicate (*Bray* and *Thomas*, 1975; *Bray* and *Thomas*, 1976; *Gallagher* et al., 1976; *Tilney*, 1976; *Weihing*, 1976) that actin in nonmuscle cells exists in two forms. *Bray* and *Thomas* (1975) estimated that in fibroblasts about half the actin was present in an unpolymerized form (G-actin) and the remaining part was filamentous and distributed largely in bundles beneath the cell cortex and within filopodia.

According to most results human or experimentally produced anti-actin sera react in immunofluorescence experiments mainly with actin distributed in microfilaments (*Lazarides* and *Weber*, 1974; *Goldman* and *Lazarides*, 1975; *Norberg* et al., 1975). *Norberg* et al. (1977) compared the staining pattern of anti-actin antibodies on various cells smeared on glass to the relative amount of cellular actin estimated by SDS gel electrophoresis with subsequent scanning of the gel. Although the cells showed a varying stainability under different conditions the cellular actin content was fairly constant. Thus, the staining differences seemed to reflect changes in the organization of cellular actin rather than actual differences in the amount of actin.

There is no definite explanation of the decreased reactivity of anti-actin antibodies with nonfilamentous actin. Recently several researchers (*Tilney*, 1976; *Kane*, 1976; *Weihing*, 1976; *Carlsson* et al., 1977) have pointed out the existence of cellular proteins that interact with actin and keep it in an unpolymerized storage form. The proteins interacting with actin might in some way interfere with the stainability of unpolymerized actin. This is supported by our experience of the influence of DNase I on the anti-actin staining of smeared lymphoblastoid cells. DNase I causes depolymerization of filamentous actin and forms complexes consisting of 1 mol of actin and 1 mol of the enzyme (*Hitchcock*, et al., 1976). The presence of myosin and tropomyosin highly influences the depolymerization of the F-actin and the formation of DNase I–actin complexes. Treatment of smeared lymphoblastoid cells with DNase I totally blocked the anti-actin staining of these cells. On the other and, the staining of cryostat sections from smooth or skeletal muscle was not influenced by DNase I treatment, nor was the staining of cells, e.g., fibroblasts, grown on glass. This might depend upon the different organization of the contractile proteins in the smeared lymphoblastoid cells and the other tissues used.

In this context it should be mentioned that human sera as well as sera from all animals examined (dog, cattle, sheep, rabbit, guinea pig, rat, and mouse) contain a substance which competes with antibodies in the same way as DNase I. Its reaction with actin is Ca^{2+}-dependent and destroyed by heating the serum to 56°C for 30 min (to be published). Therefore all anti-actin sera, human as well as animal sera, should be heated before use.

B. Comparison of Human and Rabbit Anti-Actin Antibodies

The human sera can easily be obtained in ample amounts whereas the production of experimental antisera is time consuming and highly laborious. Thus, it seems to be extremely important to characterize and compare the reactivity of anti-actin antibodies of different origin in order to find out whether they can be used alternatively or whether their reactions with actin diverge in any respect.

Both human and experimentally produced rabbit anti-actin antibodies have been used to study muscular structures. They stain the thin filaments (I-bands) of myofibrils in the same way (*Andersen* et al., 1976; *Trenchev* and *Holborow*, 1976; *Chaponnier* et al., 1977). The staining of sections of smooth muscles in, e.g., vessels, stomach and uterus is also indistinguishable (*Trenchev* et al., 1974; *Trenchev* and *Holborow*, 1976; *Fagraeus* et al., 1977). The actin-containing structures of mouse fibroblasts and rat embryo cells grown on glass have been extensively studied by *Lazarides* (1975) using IFL and rabbit antibodies against actin isolated from smooth muscle or mouse fibroblasts. In our studies using human skin fibroblasts (*Norberg* et al., 1975; *Fagraeus* et al., 1978), human anti-actin antibodies and rabbit antibodies against rabbit or pig skeletal muscle actin gave the same IFL-staining with fibers showing a continuous fluorescence along their lengths. Nor have we been able to show any difference between human and rabbit anti-actin antibodies in staining lymphoid cells and platelets (*Norberg* et al., 1977). Thus, so far human and experimentally produced rabbit anti-actin antibodies seem to be exchangeable in indirect tests, such as IFL experiments, although several results in other in vitro systems indicate that their antigen specificity might not be identical.

Most human sera containing anti-actin antibodies precipitate easily with actin in gel diffusion experiments (*Gabbiani* et al., 1973; *Lidman* et al., 1976; *Andersen* et al., 1976; *Chaponnier* et al., 1977) whereas in our experience rabbit anti-actin sera do not. We have used our own antisera against pig and rabbit skeletal muscle actin, rabbit sera against pig skeletal actin (kindly provided from Dr. M. Crumpton, MRC, Mill Hill, London), and rabbit antiserum against gizzard actin purified on SDS–polyacrylamide gel electrophoresis (a generous gift from Dr. K. Burridge, Cold Spring Harbor Laboratory). They all gave negative results in immunodiffusion experiments using pig or rabbit skeletal F-actin which had been reduced by 0.0005 M mercaptoethanol in order to make the actin diffusible. However, F-actin heated for 30 minutes at 70°C was precipitated by the rabbit anti-actin sera (*Norberg* et al., unpublished).

Other researchers (*Lazarides* and *Weber*, 1974; *Lazarides*, 1975; *Trenchev* and *Holborow*, 1976) have reported precipitin reactions of their rabbit anti-actin sera but they have not stated how the actin used as antigen was treated. Human and rabbit anti-actin sera also reacted differently in passive hemagglutination tests (*Fagraeus* et al., 1977). Rabbit sera readily agglutinated actin-coated sheep red cells whereas human anti-actin sera did not. The agglutination capacity of rabbit sera was restricted to the IgM fraction obtained by gradient centrifugation of rabbit anti-actin sera. None of human sera with anti-actin antibodies of IgG class gave a positive passive hemagglutination reaction. A further support of the differences in specificity between human and experimentally produced antisera might be deduced from the result of *Lessard* et al. (1977). They have developed a sensitive and specific radioimmunoassay to quantitate anti-actin antibodies. The method involves the detection of an immune complex between actin, covalently linked to agarose, and IgG on the basis of the specific binding of ^{125}I-labeled protein A from *Staphylococcus aureus*. The method has been used to follow the appearance of anti-actin antibodies in rabbits immunized with SDS-denatured and electrophoretically purified murine skeletal muscle actin. However, two of our human anti-actin sera and a number of SMA-positive sera from local patients with CAH failed to react in this radioimmunoassay (Lessard, personal communication).

So far, there are no definite explanations for the differing behaviors of human and experimentally produced anti-actin sera. The type and availability of the antigenic determinants of actin may vary depending on the source and on the methods used for preparation of actin, the way of immunization, whether they are spontaneously occurring or experimentally produced etc. Moreover, the test system used may influence the availability of the antigenic determinants of actin.

References

Ajdukiewicz, A.B., Dudley, F.J., Fox, R.A., Doniach, D., Sherlock, S.: Immunological studies in an epidemic of infective short-incubation hepatitis. Lancet *i*, 7755–7757 (1972)

Andersen, P.: Smooth muscle antibodies in sera with Paul-Bunnell heterophil antibodies. Acta Path. Microbiol. Scand. Sect. C *83*, 19–27 (1975)

Andersen, P., Andersen, H.K.: Smooth muscle antibodies and other tissue antibodies in cytomegalovirus infection. Clin. Exp. Immunol. *22*, 22–29 (1975)

Andersen, P., Molgaard, J., Andersen, J., Andersen, H.K.: Smooth muscle antibodies and antibodies to cytomegalovirus and Epstein-Barr virus in leukaemias and lymphomata. Acta Path. Microbiol. Scand. Sect. C *84*, 86–92 (1976)

Andersen, P., Small, J.V., Sobieszek, A.: Studies on the specificity of smooth muscle antibodies. Clin. Exp. Immunol. *26*, 57–66 (1976)

Biberfeld, G., Fagraeus, A., Lenkei, R.: Reaction of human smooth muscle antibody with thyroid cells. Clin. Exp. Immunol. *18*, 371–379 (1974)

Biberfeld, G., Sterner, G.: Smooth muscle antibodies in Mycoplasma pneumoniae infection. Clin. Exp. Immunol. *24*, 287–291 (1976)

Botazzo, G.F., Florin-Christensen, A., Fairfax, A., Swana, G., Doniach, D., Groeschel-Stewart, U.: Classification of smooth muscle autoantibodies detected by immunofluorescence. J. Clin. Path. *29*, 403–410 (1976)

Bray, D.: Anti-actin. Nature (London) *251*, 187 (1974)

Bray, D., Thomas, C.: The actin content fibroblasts. Biochem. J. *147*, 221–228 (1975)

Bray, D., Thomas, C.: Unpolymerized actin in tissue cells. In: Cell Motility. *Goldman, R., Pollard, T., Rosenbaum, J.*, (Eds.) Cold Spring Harbor Lab. 1976, p. 461–473

Carlsson, L., Nyström, L.E., Lundkvist, J., Markey, F., Lindberg, U.: Actin polymerizability is influenced by profilin, a low molecular weight protein in non-muscle cells. J. Mol. Biol. (in press)

Chaponnier, Ch., Kohler, L., Gabbiani, G.: Fixation of human anti-actin autoantibodies on skeletal muscle fibres. Clin. Exp. Immunol. *27*, 278–284 (1977)

Collins, J.H., Elzinga, M.: The primary structure of actin from rabbit sekeletal muscle. Five cyanogen bromide peptides, including the NH_2 and COOH termini. J. Biol. Chem. *250*, 5987–5998 (1975)

Elzinga, M., Collins, J.H., Kuehl, W.M., Adelstein, R.S.: Complete amino acid sequence of actin of rabbit skeletal muscle. Proc. Nat. Acad. Sci. *70*, 2687–2693 (1973)

Elzinga, M., Moran, B.J., Adelstein, R.S.: Human heart and platelet actins are products of different genes. Science *191*, 94–97 (1975)

Fagraeus, A., Biberfeld, G., Norberg, R.: Human and rabbit anti-actin antibodies. Annal. Immunol. (Inst. Pasteur) 129 C, 245–254 (1978)

Fagraeus, A., Lidman, K., Norberg, R.: Indirect immunofluorescence staining of contractile proteins in smeared cells by smooth muscle antibodies. Clin. Exp. Immunol. *20*, 467–477 (1975)

Fagraeus, A., The, H., Biberfeld, G.: Reaction of human smooth muscle antibody with thymus medullary cells. Nature London *246*, 113–115 (1973)

Fairfax, A.J., Groeschel-Stewart, U.: Myosin autoantibodies detected by immunofluorescence. Clin. Exp. Immunol. *28*, 27–34 (1977)

Farrow, L.J., Holborow, E.J., Brighton, W.D.: Reaction of human smooth muscle antibody with liver cells. Nature London *232*, 186–187 (1971)

Farrow, L.J., Holborow, E.J., Johnson, G.D., Lamb, S.G., Stewart, J.S., Taylor, P.E., Zuckerman, A.J.: Autoantibodies and the hepatitis-associated antigen in acute infective hepatitis. Br. Med. J. 2, 693–695 (1970)

Gabbiani, G., Ryan, G.B., Badonnel, M.-C., Majno, G.: Smooth muscle antigens in platelets. Immunofluorescent detection using human anti-smooth muscle serum. Path. Biol. *20*, 6–8 (1972)

Gabbiani, G., Ryan, G., Lamelin, J-P., Vassalli, P., Majno, G., Bouvier, G., Cruchaud, A., Lüscher, E.: Human smooth muscle antibody. Am. J. Pathol. *72*, 473–488 (1973)

Galbraith, R.M., Smith, M., Mackenzie, R.M., Tee, D.E., Doniach, D., Williams, R.: High prevalence of seroimmunologic abnormalities in relatives of patients with active chronic hepatitis or primary biliary cirrhosis. New Engl. J. Med. *290*, 63–69 (1974)

Gallagher, M., Ostwiler, T.C., Stracher, A.: Two forms of platelet actin that differ from skeletal muscle actin. In: Cell Mobility. *Goldman, R., Pollard, T., Rosenbaum, J.* (Eds.) Cold Spring Harbor Lab. 1976, p. 475–486

Goldman, R.D., Lazarides, E., Pollack, R., Weber, K.: The distribution of actin in non-muscle cells. Exptl. Cell Res. *90*, 333–344 (1975)

Hirabayashi, T., Hayashi, Y.: Antibody specific for actin from frog skeletal muscle. J. Biochem. *71*, 153–156 (1972)

Hitchcock, S., Carlsson, L., Lindberg, U.: Depolymerization of F-actin by deozyribonuclease I. Cell 7, 531–538 (1976)

Holborow, E.J., Hemsted, E.H., Mead, S.V.: Smooth muscle autoantibodies in infectious mononucleosis. Br. Med. J. 323–325 (1973)

Johnson, G.D., Holborow, E.J., Glynn, L.E.: Antibody to smooth muscle in patients with liver disease. Lancet ii, 878–890 (1965)

Kane, R.E.: Actin polymerization and interaction with other proteins in temperature-induced gelation of sea urchin egg extracts. J. Cell. Biol. *71*, 704–714 (1976)

Kesztyüs, L., Nikodémusz, S., Szilágyi, T.: Antigenic activity of myosin and actin. Nature London *163*, 136 (1949)

Lamb, R.A., Mahy, B.W.J., Choppin, P.W.: The synthesis of Sendai virus polypeptides in infected cells. Virology *69*, 116 (1976)

Lamelin, J-P, Williams, E.H., Souissi, T., De-Thé, G., Gabbiani, G.: Smooth muscle antibody in Burkitt's lymphoma and in nasopharyngeal carcinoma. Clin. Exp. Immunol. *28*, 157–162 (1977)

Lazarides, E.: Immunofluorescence studies on the structure of actin filaments in tissue culture cells. J. Histochem. Cytochem. *23*, 507–528 (1975)

Lazarides, E.: Actin, α-actinin, and tropomyosin interaction in the structural organization of actin filaments in nonmuscle cells. J. Cell. Biol. *68*, 202–219 (1976)

Lazarides, E., Weber, K.: Actin antibody: The specific visualization of actin filaments in non-muscle cells. Proc. Nat. Acad. Sci. *71*, 2268–2272 (1974)

Lessard, J.L., Rein, D.C., Carlton, D., Akeson, R.: A solid-phase radioimmunoassay for actin and anti-actin IgG. Submitted as abstract for the Meeting of Cell Biology. Nov. 1977

Li, A.K.C., Trenchev, P.S., Holborow, E.J., Newsome, C., Wynne, A.T.: Experimental smooth muscle antibodies. Clin. Exp. Immunol. *27*, 273–277 (1977)

Lidman, K.: Clinical diagnosis in patients with smooth muscle antibodies. Acta Med. Scand. *200*, 403–407 (1976)

Lidman, K., Biberfeld, G., Fagraeus, A., Norberg, R., Thorstenson, R., Utter, G., Carlsson, L., Luca, J., Lindberg, U.: Anti-actin specificity of human smooth muscle antibodies in chronic active hepatitis. Clin. Exp. Immunol. *24*, 266–272 (1976)

Lidman, K., Biberfeld, G., Sterner, G., Norberg, R.: Chronic active hepatitis in children. Acta Ped. Scand. *66*, 73–79 (1977)

Mackay, J.R., Morris, P.J.: Association of autoimmune active chronic hepatitis with HL-A 1,8. Lancet *ii*, 793–795 (1972)

Marshall, J.M., Holtzer, H., Jr., Finck, H., Pepe, F.: The distribution of protein antigens in striated myofibrils. Exp. Cell. Res. Suppl. *7*, 219–233 (1959)

McMillan, S.A., Haire, M.: Smooth muscle antibody in warts patients. Clin. Exp. Immunol. *21*, 339–345 (1975)

Murray-Lyon, I.M., Stern, R.B., Williams, R.: Controlled trial of prednisone and azathioprine in active chronic hepatitis. Lancet *i*, 735–737 (1973)

Norberg, R., Biberfeld, G., Fagraeus, A., Lidman, K., Thorstensson, R., Utter, G.: The reaction of cells with anti-actin sera in relation to the amount of cellular actin. Clin. Exp. Immunol. *28*, 512–516 (1977)

Norberg, R., Lidman, K., Fagraeus, A.: Reaction of human smooth muscle antibody with human platelets. Clin. Exp. Immunol. *21*, 284–288 (1975)

Norberg, R., Lidman, K., Fagraeus, A.: Effects of cytochalasin B on fibroblasts, lymphoid cells, and platelets revealed by human anti-actin antibodies. Cell *6*, 507–512 (1975)

Olsson, R., Hultén, L.: Concurrence of ulcerative colitis and chronic active hepatitis. Scand. J. Gastroenteral. *10*, 331–340 (1975)

Owaribe, K., Hatano, S.: Induction of antibody against actin from Myxomycete plasmodium and its properties. Biochemistry *14*, 3024–3029 (1975)

Paronetto, F., Gerber, M.A., Vernace, S.J.: Immunologic studies in patients with chronic active hepatitis and primary bilary cirrhosis. I. Cytotoxic activity and binding of sera to human liver cells grown in tissue culture. Proc. Soc. Exp. Biol. (N.Y.) *143*, 756–761 (1973)

Pepe, F.A.: Some aspects of the structural organization of the myofibril as revealed by antibody-staining method. J. Cell. Biol. *28*, 505–525 (1966)

Reynolds, T.B., Peters, R.L., Yamada, S.: Chronic active and lupoid hepatitis caused by a laxative, oxyphenisatin. New Engl. J. Med. *285*, 813–815 (1971)

Shu, S., Nisengard, R.J., Hale, W.L., Beutner, E.H.: Incidence and titers of antinuclear, anti-smooth muscle and other autoantibodies in blood donors. J. Lab. Clin. Med. *86*, 259–265 (1975)

Sutton, R.N.P., Emond, R.T.D., Thomas, D.B., Doniach, D.: The occurrence of autoantibodies in infectious mononucleosis. Clin. Exp. Immunol. *17*, 427–436 (1974)

Tilney, L.G.: The polymerization of actin. J. Cell. Biol. *69*, 51–72 (1976)

Trenchev, P., Holborow, E.J.: The specificity of anti-actin serum. Immunology *31*, 509–517 (1976)

Trenchev, P., Sneyd, P., Holborow, E.J.: Immunofluorescent tracing of smooth muscle contractile protein antigens other than smooth muscle. Clin. Exp. Immunol. *16*, 125–136 (1974)

Tunik, B., Holtzer, H.: The distribution of muscle antigens in contracted myofibrils determined by fluorescein-labelled antibodies. J. Biophys. Biochem. Cytol. *11*, 67–76 (1961)
Tyrell, D., Norrby, E.: Structural polypeptides of measles virus. J. Virology, in press.
Utter, G., Biberfeld, P., Fagraeus, A., Norberg, R., Thorstensson, R.: Ultrastructure on in vitro formed actin-anti-actin immune complexes. Exp. Cell Res. 1978 (in press)
Wang, E., Wolf, B.A., Lamb, R.A., Choppin, P.W., Goldberg, A.R.: The presence of actin in enveloped viruses. In: Cell Motility. *Goldman, R., Pollard, T., Rosenbaum, J.* (Eds.) Cold Spring Harbor Lab., 1976, p. 589–600
Wasserman, J., Glas, U., Blombren, H.: Antibodies in patients with carcinoma of the breast. Clin. Exp. Immunol. *19*, 417–422 (1975)
Weihing, R.R.: Membrane association and polymerization of actin. In: Cell Motility. *Goldman, R., Pollard, R., Rosenbaum, J.* (Eds.) Cold Spring Harbor Lab., 1976, p. 671–684
Whitehouse, J.M.A., Holborow, E.J.: Smooth muscle antibody in malignant disease. Br. Med. J. *4*, 511–513 (1971)
Wittingham, S., Mackey, J.R., Irwin, J.: Autoimmune hepatitis. Immunofluorescence reactions with cytoplasm of smooth muscle and renal glomerular cells. Lancet *i*, 1333–1335 (1966)
Wilson, F.J., Finck, H.: Actin: Immunochemical and immunofluorescence studies. J. Biochem. *70*, 143–148 (1971)

Structure and Replication of α-Viruses

Leevi Kääriäinen and Hans Söderlund[1]

I. Introduction

The α-virus family presently contains 20 members of closely related viruses. These viruses were previously known as group A arboviruses but are currently classified as togaviruses together with flaviviruses (previously group B arboviruses) and some nonarboviruses such as rubella and lactic dehydrogenase viruses (*Wildy,* 1971, *Fenner,* 1976). In nature, the α-viruses are spread to their mamma-

---

[1] Department of Virology, University of Helsinki, Haartmaninkatu 3, SF-00290 Helsinki 29, Finland

lian or avian hosts by arthropods, mainly mosquitoes. The viruses multiply in both vertebrate and invertebrate hosts (*Casals* and *Clarke*, 1965; *Mussgay* et al., 1975). A great deal of research has been carried out to clarify the ecology, epidemiology, and pathology of these viruses since several α-viruses are of great importance in veterinary and human medicine (*Casals* and *Clarke*, 1965; *Casals*, 1975).

This review concentrates on the molecular biology of the α-viruses, for which a huge body of literature exists. However, only 2 out of the 20 members, the nonpathogenic Semliki Forest (SF) virus and the Sindbis virus, have been extensively studied. Some data also exist on the chikungunya, eastern equine encephalitis (EEE), Venezuelan equine encephalitis (VEE), and western equine encephalitis (WEE) viruses. The viruses seem to be remarkably similar in molecular terms, and here we have often made generalizations based on the data obtained from one or two of the members, which do not of course, represent valid assumptions in all cases.

II. Structure

Production of α-virus particles on a milligram scale for morphologic, chemical, and physical studies is a relatively simple task. The virus is released from the infected cells by budding through the plasma membrane and can then be recovered from the culture medium. Several purification methods have been described, but the one generally adopted includes concentration of the virus from the culture fluid by ultracentrifugation or by precipitation, e.g., with polyethylene glycol. Purification of the concentrated preparation is then achieved by rate zonal and/or isopycnic centrifugation on sucrose or tartrate density gradients (*Kääriäinen* et al., 1969; *Strauss* et al., 1969; *David*, 1971). The purified virus particles consist of RNA, protein, protein-bound carbohydrate, and lipids (Table 1). The particle has a sedimentation constant of about 280 S and a density of 1.18–1.20 g/cm^3 in sucrose (*Cheng*, 1961; *Aaslestad* et al., 1968; *Osterrieth*, 1968; *Strauss* et al., 1968; *Horzinek* and *Mussgay*, 1969; *Kääriäinen* et al., 1969; *Fuscaldo* et al., 1971; *Mussgay* et al., 1973).

In electron micrographs, the negatively stained α-viruses appear as spheric particles of about 50–55 nm in diameter, to which 7-nm long spikes are attached

Table 1. Overall composition of α-viruses

Virus	Component (%)				Reference
	RNA	Protein	Carbohydrate (in glycoprotein)	Lipid	
Sindbis virus	5.5	61	6.5	27	*Pfefferkorn* and *Shapiro* (1974) [a]
SF virus	6.3	56.6	6.3	30.8	*Laine* et al. (1973)

[a] Data combined from *Pfefferkorn* and *Hunter* (1963a) and *Strauss* et al. (1970)

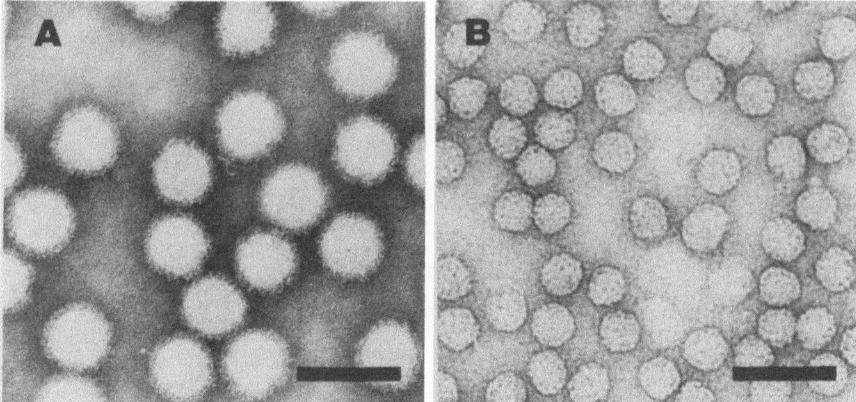

Fig. 1. Negatively stained electron micrographs of (A) SF virus particles and (B) isolated Sindbis virus nucleocapsids. Bar 100 nm (Courtesy of *Dr. C.H. von Bonsdorff*)

in radial orientation (Fig. 1). An internal core, or nucleocapsid, which is surrounded by a lipid bilayer can be visualized in thin sections.

The viral envelope can be disrupted with mild detergents and the nucleocapsid separated from the solubilized envelope components by sucrose gradient centrifugation (reviewed by *v. Bonsdorff,* 1973; *Higashi,* 1973; *Horzinek,* 1975). Treatment of the virion with strong detergents, such as sodium dodecyl sulfate, causes a total disintegration of the particle, making it possible to study the proteins and RNA (reviewed by *Pfefferkorn* and *Shapiro,* 1974; *Strauss* and *Strauss,* 1977; *Simons* et al., 1978).

A. Nucleocapsid

1. RNA

The genome of the virion consists of one continuous single-stranded RNA molecule which has a sedimentation value of 40–49 S (*Pfefferkorn* and *Shapiro,* 1974; *Strauss* and *Strauss,* 1977). Polyacrylamide gel electrophoresis has given molecular weight estimates ranging from 3.6–4.6×10^6 (*Cartwright* and *Burke,* 1970; *Dobos* and *Faulkner,* 1970b; *Arif* and *Faulkner,* 1971; *Levin* and *Friedman,* 1971; *Agabalyan* et al., 1972; *Boulton* and *Westaway,* 1972; *Simmons* and *Strauss,* 1972a). The electrophoretic mobility of double-stranded genome RNA indicates a molecular weight of 4.4×10^6 (*Simmons* and *Strauss,* 1972b) and sedimentation in sucrose gradients in the presence of formaldehyde of 4.1×10^6 (*Simmons* and *Strauss,* 1972a). Electron microscopic length measurement of glyoxal-denatured Sindbis RNA gives a value of 4.6×10^6 daltons (*Hsu* et al., 1973). The wide range is probably due to difficulties in determining the molecular weight of a large single-stranded RNA rather than to true variation. A commonly used figure is $4.3 \pm 0.2 \times 10^6$ which corresponds to about 13,000 nucleotides.

The genome RNA shares properties with eukaryotic messenger RNAs; the

3' end contains a poly (A) tract, which is on the average about 60 nucleotides long, and possibly has some heterogeneity (*Armstrong* et al., 1972; *Eaton* and *Faulkner,* 1972; *Johnston* and *Bose,* 1972; *Hsu* et al., 1973; *Wittek* et al., 1974; *Deborde* and *Leibowitz, 1976),* while the 5' end of at least the Sindbis virus RNA has an inverted 7-methylguanosine, i.e., a cap (*Hefti* et al., 1976; *Dubin* et al., 1977).

Protein-free Sindbis virus RNA has some secondary structure (*Sprecher-Goldberger,* 1967; *Engelhardt,* 1972; *Brawner* et al., 1977). This RNA can circularize as reported by *Hsu* et al. (1973), probably due to cohesive ends which form a stem, about 250 nucleotides long, in a panhandle-like structure. In spite of the close similarity between different α-viruses, their RNAs have a strikingly low base sequence homology, as revealed by cross-hybridization and oligonucleotide fingerprinting of Sindbis, SF, chikungunya, and O'nyong-nyong RNAs (*Wengler* et al., 1977).

2. Capsid Protein

In addition to the RNA, the nucleocapsid contains 200–300 capsid protein molecules, which is 66% of the particle's weight (*Strauss* et al., 1968; *Kääriäinen* et al., 1969; *Acheson* and *Tamm,* 1970b; *Igarashi* et al., 1970; *Laine* et al., 1973). The capsid protein has a molecular weight of 30,000–33,000 as determined by SDS polyacrylamide gel electrophoresis and gel filtration (*Strauss* et al., 1968; *Simons* and *Kääriäinen,* 1970; *Kennedy* and *Burke,* 1972). The molecular weight seems to be approximately the same in the six most commonly studied α-viruses (chikungunya, EEE, SF, Sindbis, VEE, and WEE) (*Pedersen* et al., 1974). The capsid proteins from different α-viruses also share common antigenic determinants (*Dalrymple* et al., 1973). The capsid protein has a high lysine content (SF virus, *Simons* and *Kääriäinen,* 1970; *Kennedy* and *Burke,* 1972) or a high lysine plus arginine content (Sinbis virus, *Burke* and *Keegstra,* 1976). A low level of phosphorylation, less than 0.05 M phosphate per mole protein, has been reported for the capsid protein of the Sindbis virus (*Waite* et al., 1974).

3. Structure of the Nucleocapsid

The nucleocapsid particle isolated from purified virions after treatment with mild detergents, or from infected cells, has been studied intensively. The sedimentation value of the particle is 140–150 S and density values of 1.40–1.47 g/cm^3 in CsCl after fixation with glutaraldehyde have been reported. (*Dobos* and *Faulkner,* 1968; *Strauss* et al., 1968; *Kääriäinen* et al., 1969; *Acheson* and *Tamm,* 1970a; *Appleyard* et al., 1970; *Horzinek* and *Mussgay,* 1971; *Uryayev* et al., 1971; *Zhdanov* et al., 1972; *Igarashi* et al., 1973; *Karabatsos,* 1973). The nucleocapsids are spheric with a diameter of 35–40 nm (*Osterrieth* and *Calberg-Bacq,* 1966; *Kääriäinen* et al., 1969; *Acheson* and *Tamm,* 1970a; *v. Bonsdorff,* 1972, 1973; *Söderlund* et al., 1972; *Horzinek,* 1975). Small angle X-ray scattering studies of whole virions have revealed diameter values of 40 nm for the Sindbis and 38 nm for the SF virus nucleocapsids (*Harrison* et al., 1971; *Harrison* and *Kääriäinen*, unpublished data). Thus, the smaller diameter values reported are probably artifacts due to a shrinkage of the nucleocapsids during the specimen preparation. However, *Brown* and *Gliedman* (1973) have reported the existence of smaller nucleocapsids together with the normal ones in persistently infected mosquito cells.

Most authors agree that the particle is icosahedral (*McGee-Russel* and *Gosztonyi*, 1967; *Faulkner* and *McGee-Russel, 1968; Osterrieth,* 1968; *Simpson* and *Hauser,* 1968; *Horzinek,* 1973 a, b). Since no staining methods have revealed distinct surface structures, the exact symmetry has not been established. Triangulation numbers of T=3 (*Horzinek* and *Mussgay,* 1969), T=4 (*v. Bonsdorff,* 1973; *Harrison* et al., 1974), and T=9 (*Brown* et al., 1972; *Brown* and *Gliedman,* 1973) have been suggested.

The protein shell of at least SF virus nucleocapsid is not closed as in most naked virions (e.g., picorna, *Rueckert,* 1976 or isometric plant viruses, *Bancroft,* 1970), since the RNA is sensitive to the action of ribonuclease (*Acheson* and *Tamm,* 1970c; *Kääriäinen* and *Söderlund,* 1971). A structural role for the RNA has been proposed because ribonuclease digestion of the SF virus nucleocapsid causes a contraction of the nucleocapsid diameter to 32 nm (*v. Bonsdorff,* 1972). Exposure of SF virus nucleocapsids to slightly acid pH (5.6–6.2) results in irreversible contraction of particles from 38 to 32 nm in diameter. The contracted particle has an increased sedimentation value of 166 *S,* but the RNA to protein ratio remains unchanged (*Söderlund* et al., 1972). The mechanism and significance of this phenomenon is unknown, but it is interesting to note that the configurational change occurs at the same pH range at which the carboxyl-carboxylate pair of tobacco mosaic virus capsid protein dissociates (*Butler,* 1971). This pair apparently controls the TMV assembly. Furthermore, the SF-virus nucleocapsid can be "unfolded" to a slowly sedimenting strand-like structure without loss of protein or RNA by treatment with EDTA in low salt. Similar structures are produced in low concentrations of sodium dodecyl sulfate (*Söderlund* and *Kääriäinen,* 1974; *Söderlund* et al., 1975). The SF virus nucleocapsid is completely dissociated into RNA and protein at 0.2 m*M* SDS. This sensitivity to SDS is indicative of strong RNA protein interactions in the particle according to *Kaper* (1973). Some properties of the SF virus nucleocapsid are summarized in Table 2.

Table 2. Properties of SF virus nucleocapsid in its different configurations

Parameter	pH 7.2	pH 6.2	EDTA	RNase	SDS 0.05 m*M*
Size (nm)	39	32	> 50	32	> 50
S value	150 *S*	166 *S*	100 *S*	100–130 *S*	100 *S*
Density (g/cm^3)	1.43	1.43	1.43	1.38–1.41	ND
RNA/protein	0.5	0.5	0.5	0.25–0.45	0.5

From *Söderlund* et al. (1975)

B. Envelope

1. Envelope Proteins

Two different envelope glycoproteins E1 and E2 with molecular weights of about 50,000 can be resolved by polyacrylamide gel electrophoresis from EEE, SF, Sindbis, VEE, and WEE viruses (*Schlesinger M.* et al., 1972; *Garoff* et al., 1974; *Ivanic,* 1974; *Pedersen* et al., 1974). Since the mobility of the two proteins in polyacrylamide gels can very greatly between different viruses, it has been suggested that E2 should be defined as a protein which is formed by cleavage of a precursor protein p62 (or pE_2) (see Sect. III. D. 1). This nomenclature [(recommendation of the International Arbovirus Meeting in Helsinki 1975, J. Gen. Virol. 30, 273 (1976)] is adopted here. In a comparative study by *Pedersen* et al. (1974), the chikungunya virus failed to give two envelope protein bands. This is apparently due to difficulties in separation rather than real qualitative differences. SF virus has an additional small, highly glycosylated protein E3 with a molecular weight of about 10,000 (*Garoff* et al., 1974), first found in infected cells treated with canavanine (*Ranki* et al., 1972). This protein has not yet been reported in other α-viruses.

The envelope proteins of the SF and Sindbis viruses have been purified by chromatography on hydroxylapatite in the presence of SDS or on DEAE cellulose in the presence of triton (*Garoff* et al., 1974; *Burke* and *Keegstra,* 1976). Sindbis virus E1 and E2 separate at pI 6 and pI 9, respectively, in preparative isoelectric focusing, carried out in the presence of triton (*Dalrymple* et al., 1976). The three envelope proteins of the SF virus can be separated by sucrose gradient centrifugation in the presence of sodium deoxycholate (*Helenius* et al., 1976).

The amino acid compositions of the envelope proteins do not show higher proportion of hydrophobic amino acids than ordinary soluble proteins (*Garoff* et al., 1974; *Burke* and *Keegstra,* 1976). Their amphiphilic nature can, however, be detected by their affinity to detergents like triton and deoxycholate which are known to bind to hydrophobic regions of integral membrane proteins (*Helenius* and *Simons,* 1972; *Uterman* and *Simons,* 1974; *Becker* et al., 1975; *Helenius* and *Simons,* 1975).

The hemagglutinating activity resides in E1 in both the Sindbis and SF viruses (*Dalrymple* et al., 1976; *Helenius* et al., 1976). The hemagglutinin also seems to be only one of the envelope proteins in the VEE virus (*Pedersen* and *Eddy,* 1974). Sindbis E1 protein carries antigenic determinants which cross-react with related α-virus antibodies while E2 is antigenically distinct (*Dalrymple* et al., 1976). Strain specificity of the envelope proteins is also shown by differences in their electrophoretic mobility (*Pedersen* et al., 1974; *Pedersen* and *Eddy,* 1975). The phosphokinase activity reported to be found in the intact virion has not been localized (*Tan* and *Sokol,* 1974; *Tan,* 1975).

The carbohydrate moieties of the envelope proteins have been characterized fairly well. Glycopeptides have been isolated after pronase digestion and their sugar composition determined (Table 3) (*Burge* and *Huang,* 1970; *Burge* and *Strauss,* 1970; *Strauss* et al., 1970; *Sefton* and *Keegstra,* 1974; *Keegstra* et al.,

Table 3. Carbohydrate composition of SF virus oligosaccharides

Glyco-protein	Type of oligo-saccharide	No. of chains per protein	Moles of carbohydrate per oligosaccharide				
			Sialic acid	Galac-tose	Fucose	Man-nose	N-acetyl glucos-amine
E1	A	1 (–2)	2	2	1	3	4
E2	A	1 (–2)	2	2	1	3	4
E2	B	3	–	–	–	5–7	2
E3	A	1 (–2)	3	3	1	3–4	5–6

Data collected from *Mattila* et al., 1976; *Pesonen* and *Renkonen*, 1976; *Pesonen* and *Renkonen*, 1977

1975; *Burke* and *Keegstra*, 1976; *Mattila* et al., 1976). The Sindbis virus E1 and E2 both contain one A-type and one B-type oligosaccharide chain (nomenclature of *Johnson* and *Clamp*, 1971), while the SF virus contains one to two A-type chains in E1 and E3 and one A-type and two to three B-type chains in E2 (Table 3). The previously reported uncharacterized chain X in the E2 (*Mattila* et al., 1976; *Kääriäinen* and *Renkonen*, 1977) is probably an A-type chain (*Pesonen*, personal communication).

Sequential degradation of the oligosaccharide chains with exo- and endoglycosidases has been carried out and the structures for the A chains of the SF virus have been deduced (*Pesonen* and *Renkonen*, 1976; *Renkonen* et al., 1976; *Haahtela* and *Renkonen*, 1978). Typical N-glycosidic A-type chains similar to those of soluble serum glycoproteins appear to be present (cf. *Spiro*, 1973). The A chains carried by the individual proteins appear to be mixtures of several glycans, and the average molecular weight of the oligosaccharides carried by E3 is distinctly larger than that of E1 glycans, which in turn may be a little larger than the average E2 oligosaccharides (*Rasilo* and *Renkonen*, 1978). The largest glycans of the SF virus, those of E3, are believed to be the most exposed ones in the intact virion, because they react most readily with sialidase and galactose oxidase (*Luukkonen* et al., 1977a).

Virions which lack the terminal sialic acids in their envelope proteins, either after enzymic removal (*Kennedy, 1974*) or when grown in *Aedes albopictus* cells (*Stollar* et al., 1976), retain their infectivity and hemagglutinating activity. Even the terminal glucosamine and galactose residues may be lacking without decreased biologic activity, suggesting that the glucoseamine-mannose core of the A-type oligosaccharide is sufficient for these activities (*Schlesinger S.* et al., 1976). For the Sindbis virus envelope proteins, *Waite* et al. (1974) have reported a low level of phosphorylation, while *Pinter* and *Compans* (1975) have shown that radiolabeled sulfate is incorporated into Sindbis virus envelope proteins.

2. Lipids

About 30% of the virion or 37% of the envelope consists of lipids (*Pfefferkorn* and *Hunter*, 1963a; *Laine* et al., 1973). The main components are phospholipid and cholesterol. The lipids are derived from the host cell (*Pfefferkorn* and *Hunter*, 1963 b), and the composition resembles that of the plasma membrane of the host (*Renkonen* et al., 1971; *Renkonen* et al., 1972 a, b). This is also true for the fatty acid composition within the different phospholipid classes (*Laine* et al., 1972) and for the glycolipids (*Renkonen* et al., 1971; *Hirshberg* and *Robbins*, 1974). The viral envelope however contains, more cholesterol compared to phospholipid than the host (*Renkonen* et al., 1971). Because of the host influence, the lipid composition in virus preparations propagated in mammalian or mosquito cells, for example, may vary considerably without other obvious alterations in virus structure (*Luukkonen* et al., 1976; *Luukkonen* et al., 1977 b, *Heydrick* et al., 1971). However, an abnormal lipid composition may also destabilize the virion (*Sly* et al., 1976). *Friedman* and *Pastan* (1969) have reported that about 60% of the phospholipids in the virion can be digested with phospholipase C without loss of infectivity.

3. Structure of the Envelope

The envelope of the α-viruses consists of a lipid bilayer with two to three proteins anchored to it. In many respects, this viral membrane resembles cellular membranes, but it is simpler in protein composition, homogenous, and easy to purify. Thus, it has been extensively studied as a membrane model (*Simons* et al., 1974; 1977; 1978; *Kääriäinen* and *Renkonen*, 1977).

The action of detergents on the SF virus has been studied using triton X-100, sodium dodecyl sulfate, and sodium deoxycholate. The solubilization process begins with the binding of the detergent to the virus, then it proceeds with increasing detergent concentration to lysis of the membrane, solubilization into lipid-protein-detergent complexes, and terminates with complete delipidization of the proteins (*Helenius* and *Söderlund*, 1973; *Becker* et al., 1975; *Helenius* et al., 1976).

The mode of attachment of the envelope proteins to the membrane and the interaction with the nucleocapsid have attracted a great deal of interest. Several lines of evidence indicate that the bulk of the proteins are located on the surface of the virion. A hydrophobic tail of E1 and E2 is embedded in the lipid bilayer, anchoring the proteins, and at least one of them penetrates the membrane, making contact with the capsid protein.

The external localization of the proteins has been shown by surface labeling techniques (*Gahmberg* et al., 1972 a; *Sefton* et al., 1973), by the accessibility of the proteins to proteolytic and glycolytic enzymes (*Osterrieth*, 1965; *Calberg-Bacq* and *Osterrieth*, 1966; *Compans*, 1971; *Ravid* and *Goldblum*, 1973; *Kennedy*, 1974), and also by the agglutinability of the virus with concanavaline A (*Oram* et al., 1971; *Birdwell* and *Strauss*, 1973). The different proteins and glycans react with varying efficiency, probably reflecting the organization of the viral surface. From the envelope proteins of the SF virus, the E3 is labeled most efficiently after treatment with galactose oxidase followed by reduction with tritiated borohydride (*Luukkonen* et al., 1977 a). The A-type oligosaccharide in Sindbis virus E2 is the only glycan, which can be completely removed by treatment with glycosidases (*McGarthy* and *Harrison*, 1977).

Treatment of the SF virus with thermolysin or subtilisin results in the formation of spike-less particles from which the glycoproteins have been removed. A short membrane protein fragment is, however, protected from proteolysis in the membrane (*Gahmberg* et al., 1972b). This hydrophobic fragment of about 5,000 daltons is very rich in hydrophobic amino acids. Both E1 and E2 contain such a peptide (*Uterman* and *Simons,* 1974), while E3 is probably not an integral membrane protein (*Simons* et al., 1978). These hydrophobic fragments apparently give the membrane proteins their tendency to aggregate into star-shaped oligomers or "rosettes" (*Simons* et al., 1973a; *Helenius* and *Bonsdorff,* 1976).

In a recent experiment, *Garoff* and *Söderlund* (1978), by synchronization of protein synthesis with high salt, produced virions which contain an increasing or decreasing specific activity gradient of ^{35}S-methionine from the N-terminal of the capsid protein toward the C-terminal of E1 (cf. Sect. III. D. 1.). By this approach, it was possible to show that the hydrophobic fragments are located at the carboxy-terminus of both E1 and E2.

When the virus is treated with protein cross-linking reagents, dimers consisting of E1 and E2 are preferentially formed (*Garoff,* 1974; *Simons* et al., 1978). Immunoprecipitation with specific antisera of such cross-linked material suggests that the spike structure of the SF virus is a trimer containing one each of E1, E2, and E3 (*Ziemiecki* and *Garoff,* 1977).

Only a small proportion, less than 10%, of the lipid bilayer is occupied by the penetrating hydrophobic fragment as shown by small angle x-ray scattering (*Harrison* et·al., 1971; 1974). The protein part affects the organization of the lipids in the bilayer, since the microviscocity is higher in the presence of envelope proteins or hydrophobic fragments than in liposomes or cellular membranes (*Sefton* and *Gaffney,* 1974; *Hughes* and *Pedersen,* 1975; *Moore* et al., 1976). It should be noted that because of the small radius of the virus the outer leaflet of the lipid bilayer contains 50% more lipid than the inner one (*Harrison* et al., 1971). This may cause some stress to the membrane since the even higher curvature of the small Sindbis virus particles formed in persistently infected mosquito cells seems to significantly destabilize the envelope (*Brown* and *Gliedman,* 1973). The lipids are at least partly assymmetrically distributed in the membrane since the glycolipids are found in the outer leaflet (*Stoffel* and *Sorgo,* 1976).

C. Virion Structure

Electron micrographs of α-viruses show spheric particles with a diameter of about 50 nm, which are surrounded by the spike layer. Electron microscopy and small angle x-ray scattering give the dimensions of the nucleocapsid, the lipid bilayer, and the spike layer *(Cheng,* 1961; *Morgan* et al., 1961; *Mussgay* and *Weibel,* 1963; *Mussgay* and *Rott,* 1964; *Chain* et al., 1966; *Osterrieth* and *Calberg-Bacq,* 1966; *Acheson* and *Tamm,* 1967; *Erlandson* et al., 1967; *Higashi* et al., 1967; *Faulkner* and *McGee-Russel,* 1968; *Osterrieth,* 1968; *Simpson* and *Hauser,* 1968; *Bykovsky* et al., 1969; *Lascano* et al., 1969; *Harrison* et al., 1971;

Brown et al., 1972; *v. Bonsdorff,* 1973; *Harrison* et al., 1974; *v. Bonsdorff* and *Harrison,* 1975) as shown in Figure 2.

Analysis of the surface structure of negatively stained Sindbis virus has revealed that the glycoproteins are organized with trimer clustering in a T=4 icosahedral surface lattice (*v. Bonsdorff* and *Harrison,* 1975) (Fig. 3). This highly organized surface structure explains why the pelleted virus is found in crystalline lattices (*Wiley* and *v. Bonsdorff,* 1978) and why the virus crystalizes from solution under suitable conditions (*Simons* et al., 1978). In thin sections of infected cells, crystalline arrays of extracellular virus have also been observed (*Higashi,* 1966; 1973).

The fact that the SF virus envelope proteins can be cross-linked to the nucleocapsid by dimethyl suberimidate indicates that at least one of the envelope proteins spans the lipid bilayer and is probably in direct contact with the underlying nucleocapsid (*Garoff* and *Simons,* 1974). Additional evidence for spanning of the envelope proteins has been obtained by labeling with formyl-^{35}S-methionyl sulfone-methyl phosphate of triton-solubilized proteins (*Gahmberg* et al., 1972a) or nucleocapsid-free membranes (*Garoff* and *Simons,* 1974). Under these conditions, new labeled oligopeptides appear which are probably derived from E2 (*Simons* et al., 1978). Treatment of the Sindbis virus with formaldehyde results in "cross-linking" of envelope proteins to the nucleocapsid, suggesting similar interaction of envelope and nucleocapsid in this virus (*Brown* et al., 1973).

It seems logical to assume that the symmetric arrangement of the glycoproteins in the membrane would be dictated by the direct interaction with the icosahedral nucleocapsid. This would mean that the proteins of the nucleocapsid are also organized in a T=4 surface lattice. Such an exactly defined particle should contain 240 copies of each of the structural proteins, assuming that one capsid protein interacts with one envelope protein trimer (E1-E2-E3, SF virus) or dimer (E1-E2, Sindbis virus). This model defines the number of polypeptides and the total amount of protein in the virus particles. Setting the molecular weight of the RNA at 4.3×10^6, the amount of phospholipids can be estimated from direct chemical measurements of total phosphorus in the virus compared to lipid and RNA phosphorus (*Laine* et al., 1973) or distribution of radioactive phosphorus between RNA and phospholipids in ^{32}P equilibrium-labeled SF virus particles (*Luukkonen* et al., 1976).

The values obtained with both methods agree well, giving an RNA to phospholipid ratio of 0.82. This would give 16,000–17,000 phospholipid-cholesterol pairs for the whole virus particle; hence the total molecular weight of SF virus would be close to 60×10^6 (Table 4).

If the chemical composition of this idealized particle is calculated back and compared to that based on direct measurements, there is an obvious discrepancy (Table 4 B). The reason for this is not known, and further studies are needed to establish the detailed molecular structure of the α-viruses. Among these, direct molecular weight determination of the whole particle would be essential. Part of the discrepancy must lie in the inherent difficulty of determining the exact molecular weights of the envelope glycoproteins (*Garoff* et al., 1974).

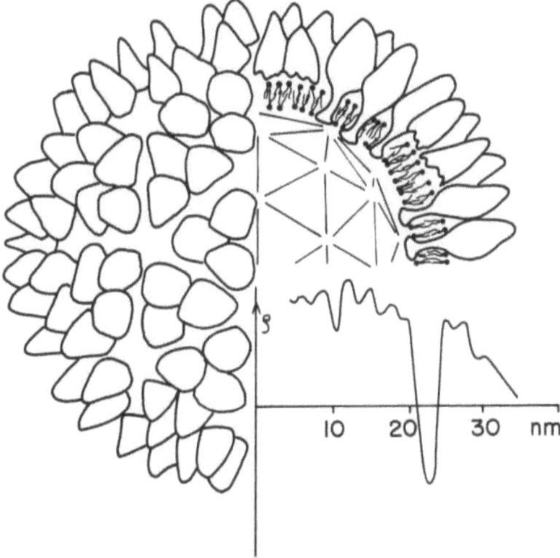

Fig. 2. A schematic drawing (modified from *McCarthy* and *Harrison* 1977) of an idealized α-virus particle showing the trimer clustering of the envelope glycoproteins in a T = 4 icosahedral surface lattice. Each subunit at the surface consists of glycoproteins E1 and E2 (plus E3 in SF virus). The surface of the nucleocapsid is drawn according to T = 4 symmetry. The inset shows the relative electron density (in arbitrary units) obtained by small angle x-ray scattering of the SF virus (*Harrison* and *Määriäinen,* unpublished results). The deep minimum corresponds to the hydrocarbon region of the lipid bilayer

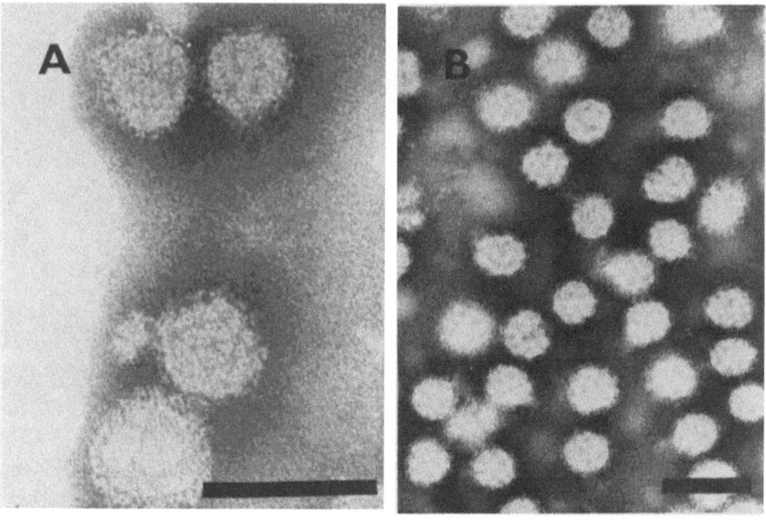

Fig. 3. Electron micrographs of negatively stained (A) Sindbis and (B) SF virus particles showing regular surface lattice. Bar 100 nm (Courtesy of *Dr. C.H. von Bonsdorff*)

Table 4A. The molecular composition of α-viruses as deduced from the T=4 symmetry model

Component	Mol wt of component	Molecules per virion	Total mol wt $\times 10^{-6}$
Nucleocapsid			11.3–12.4
RNA	$4.1–4.5 \times 10^6$	1	4.1–4.5
Protein (C)	$30–33 \times 10^3$	240	7.2–7.9
Envelope			46.0–47.2
Proteins			23.8–26.9
E1	$49–50 \times 10^3$	240	11.8–12.0
E2	$50–52 \times 10^3$	240	12.0–12.5
(E3)	10×10^3	240	2.4
Lipids			19.6–20.8
Phospholipids	775	16,000–17,000	12.3–13.1
Cholesterol	385	16,000–17,000	6.1–6.5
Glycolipids	1200	1000	1.2
Virion			54.7–60.1

The number of each component was deduced as described in the text

Table 4B. The chemical composition of α-viruses

	Dry weight %				
	RNA	Capsid protein	Envelope glycoprotein	Lipid	Particle weight $\times 10^{-6}$
Determined [a]	6.3	12.2	50.7	30.8	68 [c]
Deduced [b]	7.3	12.9	45.3	34.5	58.6

[a] *Laine* et al. (1973)
[b] From Table 4A, using the mean values when ranges are given, E3 included
[c] Based on RNA mol wt 4.3×10^6

III. Replication

A. Growth

The α-viruses grow in the cytoplasm of a large variety of vertebrate and invertebrate cell cultures in a wide temperature range between 20° and 41° C (*Pfefferkorn* and *Shapiro*, 1974; *Follet* et al., 1975; *Strauss* and *Strauss*, 1977). At 37° C, the one-step growth curve (Fig. 4) is completed usually after 6–10 h (*Dulbecco* and *Vogt*, 1954; *Rubin* et al., 1955; *Veckenstedt* and *Wagner*, 1973), whereas at 27°–29° C, the time is about twice as long (*Burge* and *Pfefferkorn*, 1966a; *Tan* et al., 1969; *Davey* et al., 1973; *Atkins* et al., 1974; *Keränen* and *Kääriäinen*, 1974; *Renz* and *Brown*, 1976). In most cases, the multiplication of α-virus usually causes severe cytopathic changes and the host cell dies within 10–20 h at 37° C (*Hardy* and *Brown*, 1961; *Acheson* and *Tamm*, 1967; *Erlandson*

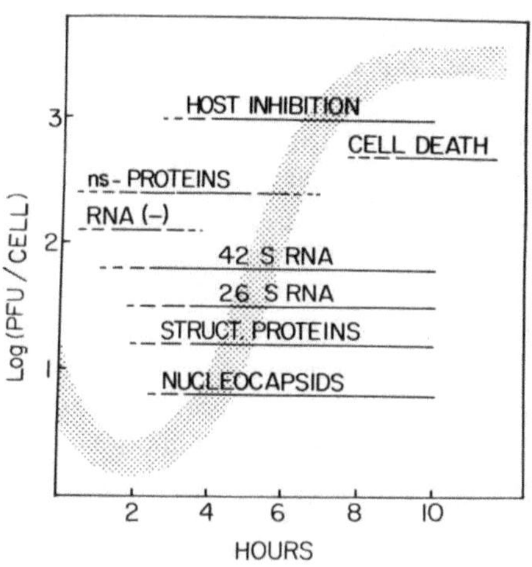

Fig. 4. One-step growth curve of the SF virus in cell culture. Schematic presentation of the duration of the major intracellular events during the virus replication

et al., 1967). The host cell protein, RNA and DNA syntheses, are inhibited within 3–5 h after infection (*Taylor*, 1965; *Lust*, 1966; *Strauss* et al., 1969; *Mussgay* et al., 1970; *Mantani* and *Kato*, 1975; *Atkins*, 1976; *Simizu* et al., 1976) by mechanisms which are still largely unknown. Persistent infections in vertebrate cells have also been described (*Schwoebel* and *Ahl*, 1972; *Inglot* et al., 1973; *Precious* et al., 1974; *Seamer*, 1974; *Schwoebel* et al., 1975; *Eaton* and *Hapel*, 1976). In the cytolytic infection, up to 20,000 progeny particles are released from the cells into the medium (*Pfefferkorn* and *Shapiro*, 1974; *Tuomi* et al., 1975; *Strauss* and *Strauss*, 1977), representing in some cases close to 10% of the phospholipid mass of the host cell plasma membrane (*Brotherus* and *Renkonen*, 1977; *Tuomi* et al., 1975). Growth in cultivated invertebrate cells at 28° C takes place more slowly and often without cytopathic effect and inhibition of host macromolecular syntheses (*Mims* et al., 1966; *Singh* and *Paul*, 1968; *Buckley*, 1969; *Peleg*, 1969; *Stevens*, 1970; *Peleg*, 1972; *Davey* et al., 1973; *Raghow* et al., 1973; *Gliedman* et al., 1975). Infection becomes limited, even if all cells are infected at the start, leading to persistent infection with a small percentage of cells producing virus (*Davey* and *Dalgarno*, 1974; *Bras-Herreng*, 1975; *Esparza* and *Sanches*, 1975; *Schlesinger R.W.*, 1975). Appearance of temperature-sensitive mutants from persistently infected mosquito cells has also been described (*Stollar* et al., 1973; *Shenk* et al., 1974; *Igarashi* et al., 1977. *Sarver* and *Stollar* (1977) have reported cytolytic infection of cloned mosquito cells.

In the following discussion, we will concentrate mainly on the cytolytic infection of vertebrate cells, the biochemical events of which are well-known (see also the excellent reviews by *Pfefferkorn* and *Shapiro*, 1974 and *Strauss* and *Strauss*, 1977).

B. Early Events

The adsorption of WEE virus to chick embryo fibroblasts was shown to be rapid and effective, more than 80% of the infectious virus being adsorbed within 30 min (*Dulbecco* and *Vogt*, 1954). Smaller percentages (20%–50%) have been shown to adsorb when radioactive virus has been used (*Birdwell* and *Strauss*, 1974b; *Marker* et al., 1977). This may be due to the presence of noninfectious, labeled particles which are unable to adsorb to the cells. There are about 10^5 receptor sites for the Sindbis virus in both chick and BHK 21 cells. These receptors are distributed evenly on the cell surface provided that the cells are fixed prior to adsorption of the virus (*Birdwell* and *Strauss*, 1974b). The nature of the receptors is still unknown. Different lipid receptors have been suggested by several investigators on the basis of the ability of different lipids to inhibit the hemagglutination of the virus (*Quersin-Thiry*, 1961; *Quersin-Thiry* and *Nihoul*, 1961; *Salminen*, 1962; *Nicoli*, 1965; *v. Frish-Niggemayer*, 1967; *Gorman*, 1970). Binding of the Sindbis virus to liposomal model membranes has been demonstrated by *Mooney* et al. (1975). The binding of radiolabeled Sindbis virus was almost quantitative to liposomes consisting of phosphatidyl ethanolamine and cholesterol. The optimum pH range was, however, from pH 6 to pH 3, suggesting that this model system does not reflect the physiologic adsorption.

Adsorption of the Sindbis virus to chick cells is dependent on the ionic strength of the medium (*Pierce* et al., 1974). More virus is bound to the cells at low ionic strength but about half of this is loosely bound and can be detached by treatment with 0.25 M NaCl. Virus binds tightly at an ionic strength of 0.15–0.17 M salt and cannot be washed away. An optimum pH of close to 6.5 has been reported for adsorption of some α-viruses (*Negro-Ponzi*, 1967; *Marker* et al., 1977).

Removal of the virion glycoprotein spikes by proteolytic enzymes renders the virus inactive (*Osterrieth*, 1965; *Compans*, 1971; *Sefton* and *Gaffney*, 1974; *Uterman* and *Simons*, 1974), indicating that the envelope proteins are responsible for the attachment of virus to the cells. It has also been shown that the membrane-free nucleocapsid is noninfectious under normal assay conditions (*Sreevalsan* and *Allen*, 1968; *Bose* and *Sagik*, 1970; *Dobos* and *Faulkner*, 1970a). Envelope protein E1 is responsible for the hemagglutinating activity of the Sindbis (*Dalrymple* et al., 1976) and SF viruses (*Helenius* and *v. Bonsdorff*, 1976), whereas E2 protein of the Sindbis virus elicited antibodies which specifically neutralized only this virus (*Dalrymple* et al., 1976). The recent results by *Burge* indicate that Sindbis virus E2 protein has hemolytic activity and may thus be responsible for the possible fusion with host cell plasma membrane (*Burge*, personal communication; see also *Ueba* and *Kimura*, 1977). The exact roles of the individual proteins, however, remain to be determined, for example, by competition experiments with isolated proteins. The subsequent steps of virus entry are poorly understood. Does the viral envelope fuse with the plasma membrane (*Morgan* and *Howe*, 1968) or is the virus taken into the cell by pinocytosis? Serious attempts to solve this question have not yet been made.

α-Viruses were among the first which were shown to have infectious RNA (*Wecker* and *Schäfer*, 1957; *Cheng*, 1958; *Wecker*, 1959a, b; *Sonnabend* et al., 1967; *Yoshinaka* and *Hotta*, 1971). This observation proved that the parental viral proteins are not needed for the initiation of the α-virus replication process. It was soon shown that protein synthesis was required during the early phase of infection (*Wecker* et al., 1962; *Wecker*, 1963; *Sreevalsan* and *Lockart*, 1964). These results indicated that the parental RNA of α-viruses served as a messenger for the synthesis of their own RNA polymerase, which then starts the replication of RNA. Support for this hypothesis was obtained from experiments with interferon, which inhibits the multiplication of α-viruses (*Friedman* and *Sonnabend*, 1965). In interferon-treated cells, the parental RNA was not converted into a double-stranded form (*Friedman* et al., 1967; *Pfefferkorn* et al., 1967), which was considered to be the first sign of the newly formed RNA polymerase activity (*Burge* and *Pfefferkorn*, 1967; *Pfefferkorn* and *Burge*, 1967; 1969). The parental RNA associates with membrane structures (*Friedman* and *Sreevalsan*, 1970). Protein synthesis may be required for this association to take place, since cycloheximide can inhibit it. In any case, RNA polymerase activity is not needed for the binding since parental RNA of temperature-sensitive RNA-negative mutants of the Sindbis virus do attach to the intracellular membranes (*Sreevalsan*, 1970).

C. α-Virus-Directed RNA Synthesis

1. α-Virus-Specific RNAs

Labeling of α-virus-infected cells with ^{3}H-uridine in the presence of actinomycin D reveals cytoplasmic RNA species with sedimentation values of 42 *S*, 26 *S*, and 18–22 *S* (*Sonnabend* et al., 1964; *Friedman* et al., 1966; *Friedman* and *Berezesky*, 1967; *Sreevalsan* and *Lockart*, 1966; *Sreevalsan* et al., 1968; *Poiree* et al., 1972; *Lubiniecki* and *Henry*, 1974). The 42 *S* and 26 *S* RNAs are single-stranded and only the former is infectious (*Sonnabend* et al., 1967). Some of the material sedimenting at about 20 *S* is RNAse-resistant double-stranded RNA. Short pulses of ^{3}H-uridine have revealed a heterogeneously sedimenting (20–30 *S*), partially RNAse-resistant structure which consists of nascent single strands and a double-stranded core (*Sreevalsan* and *Lockart*, 1966; *Friedman*, 1968b; *Kääriäinen* and *Gomatos*, 1969; *Stern* and *Friedman*, 1969; *Cartwright* and *Burke*, 1970; *Scholtissek* et al., 1972; *Simmons* and *Strauss*, 1972b; *Segal* and *Sreevalsan*, 1974; *Bruton* and *Kennedy*, 1975). This structure has been designated as replicative intermediate in analogy with the corresponding structure found from poliovirus-infected cells (*Baltimore* and *Girard*, 1966).

A genuine double-stranded RNA can be isolated from the α-virus-infected cells (*Kääriäinen* and *Gomatos*, 1969; *Stollar* and *Stollar*, 1970; *Stollar* et al., 1972; *Yoshinaka* and *Hotta*, 1972; *Martin* and *Burke*, 1974), which may represent a dead end in the RNA replication since it contains continuous 42 *S* RNA, which is infectious (*Wengler* et al., 1976). In addition to the above-listed RNAs, minor amounts of single-stranded 38 *S*, 33 *S*, and 20–22 *S* RNA have been found from α-virus-infected cells (*Kääriäinen* and *Gomatos*, 1969; *Levin* and *Friedman*, 1971; *Kennedy*, 1972; *Martin* and *Burke*, 1974; *Simmons* and *Strauss*, 1974a; *Colbere* and *Hannoun*, 1974; *Kennedy*, 1976). The puzzling phenomenology of the α-virus RNA pattern has been clarified by the investigations of *Simmons* and *Strauss* (1972a, 1974a), *Wengler* and *Wengler* (1976a), and *Kennedy* (1976).

The virion 42 *S* RNA and the intracellular 26 *S* RNA are both of positive polarity, and the 26 *S* RNA is identical to about one-third of the 42 *S* RNA as shown by hybridization competition experiments (*Simmons* and *Strauss*, 1972a). Oligonucleotide analysis has shown that 26 *S* RNA is identical to the 3′ end of the 42 *S* RNA (*Kennedy*, 1976; *Wengler* and *Wengler*, 1976a). The 38 *S* and 33 *S* RNAs are probably conformational variants of 42 *S* and 26 *S* RNAs, respectively (*Simmons* and *Strauss*, 1974a; *Kennedy*, 1976).

As is the case in 42 *S* RNA, the 3′ end of 26 *S* RNA has a polyadenylic acid tract of 60–70 residues long (*Eaton* et al., 1972; *Clegg* and *Kennedy*, 1974a). Shorter poly (A) tracts of 30–40 residues have also been reported (*Wittek* et al., 1974; *Deborde* and *Leibowitz*, 1976). The 5′ terminus of 26 *S* RNA is blocked by a cap, mG (5′) pppAp, which is methylated as $m_2^{2,7}G$ or $m_3^{2,2,7}G$ (*Dubin* and *Stollar*, 1975; *Hsuchen* and *Dubin*, 1976; *Dubin* et al., 1977). Internal methylation in m^5C of 26 *S* but not 42 *S* RNA has been reported by the above authors.

2. Synthesis of 42 *S* and 26 *S* RNAs

Some early experiments suggested that the 26 *S* RNA is produced by a specific cleavage from the 42 *S* RNA (*Sreevalsan* et al., 1968; *Dobos* and *Faulkner*, 1969; *Dobos* et al., 1971) or that 42 *S* RNA is formed by ligase enzymes from 26 *S* and 33 *S* RNA (*Michel* and *Gomatos*, 1973) or from smaller pieces (*Koblet* et al., 1974). These possibilities have been disproved by pulse-chase experiments, using glucosamine after the pulse to bind the excess of labeled ³H-UTP (*Scholtissek* et al., 1972; *Keränen*, 1977c). In these experiments, no interconversion of 42 *S* and 26 *S* RNA was observed. The finding of continuous 42 *S* RNA-negative strands in replicative intermediate structures and the lack of evidence of 26 *S* RNA-negative strands have established the mode of transcription in α-virus-infected cells (*Simmons* and *Strauss*, 1972b; *Bruton* and *Kennedy*, 1975; *Sawicki* and *Gomatos*, 1976). Analysis of the "replicative forms" obtained after mild RNAse treatment shows the presence of three different double-stranded cores, RF I (mol wt 8.8×10^6), RF II (mol wt 5.6×10^6), and RF III (mol wt 3.2×10^6), which represent double-stranded forms of 42 *S* RNA, a 2.8×10^6 dalton RNA, and 26 *S* RNA, respectively. The RFs are considered to be derived from two different replicative intermediates RIa and RIb, which synthesize 42 *S* and 26 *S* RNA, respectively (*Simmons* and *Strauss*, 1972b) (Fig. 5, a, b). Short pulses with ³H-uridine given in the middle of the infectious cycle label only RF I and RF III in their nascent positive strands, indicating that 42 *S* and 26 *S* RNA are synthesized with similar kinetics. Pulses of 40 min are required to fully label the RF II, showing that transcription in this part takes place, but very slowly compared to the 26 *S* RNA synthesizing part of the RIb (Fig. 5a). Yet no 2.8×10^6 daltons single-stranded RNA is released from the RIb. The poly (A) tract of both 42 *S* and 26 *S* RNA is transcribed from a poly (U) sequence with an average length of about 60 residues at the 5′ end of the 42 *S* RNA-negative strand (*Sawicki* and *Gomatos*, 1976; see also *Bruton* and *Kennedy*, 1975; *Pisano* et al., 1975; *Margotat* et al., 1976; *Wittek* et al., 1977).

The maximum rate of synthesis of 42 *S* RNA-negative strands is reached

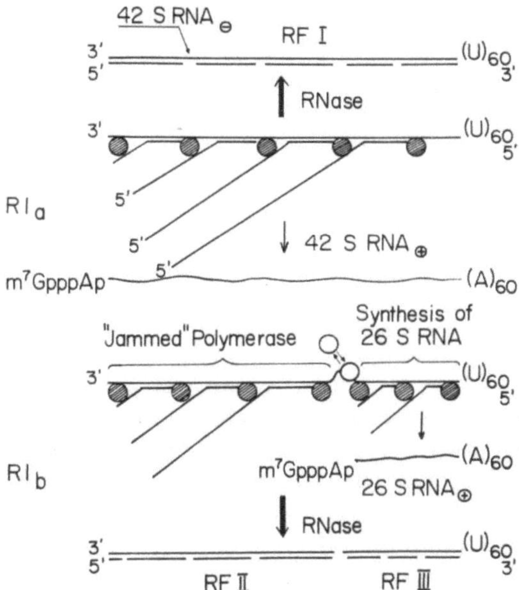

Fig. 5A. Schematic presentation of the transcription of α-virus-specific 42 *S* and 26 *S* RNA-positive strands from the 42 *S* RNA-negative strand template. The evidence for the existence of two different replicative intermediates, RIa and RIb, is based on their response to mild ribonuclease treatment which yields RF I or RF II and RF III. Interaction between interconversion protein (o) and the replication complex synthesizing 42 *S* RNA converts RIa into RIb, allowing an internal initiation of 26 *S* RNA synthesis. Proximal to this initiation site, a ribonuclease-sensitive site is created. Polymerase molecules (⊘) which started the transcription of 42 *S* RNA-positive strands prior to the action of the interconversion protein cannot continue their transcription over the initiation site of 26 *S* RNA synthesis and become "jammed." When the interconversion protein detaches, RIb is converted into RIa, and the jammed polymerase molecules can complete the interrupted transcription of the 42 *S* RNA-positive strands

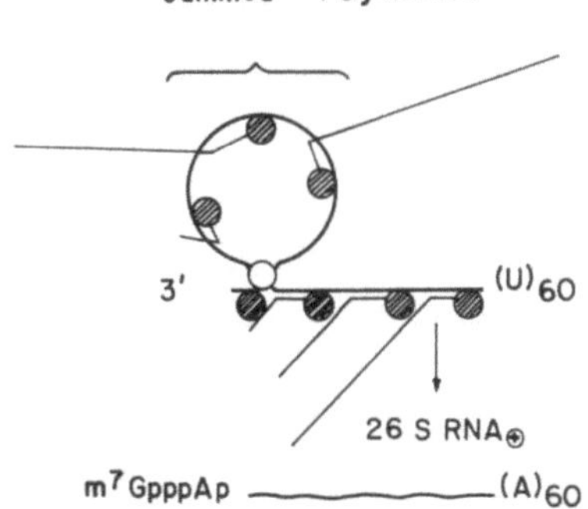

Fig. 5B. An alternative model for the action of interconversion protein. In this model, a common initiation site for transcription of 42 *S* and 26 *S* RNA is postulated. Here the interconversion protein creates a loop corresponding to RF II

at about $2^{1}/_{2}$ h postinfection, declining rapidly thereafter. The rate of synthesis of positive strands, i.e., 42 *S* and 26 *S* RNAs, grows exponentially up to 3 h postinfection remaining constant thereafter up to 6–7 h postinfection. The amount of 42 *S* and 26 *S* RNA molecules can reach values close to 200,000 per cell at 8 h postinfection as determined by equilibrium labeling with ^{32}P orthophosphate (*Tuomi* et al., 1975). The massive RNA synthesis is paralleled by a significant decrease in the nucleotide pool size of the infected cells (*Hammer* et al., 1976).

3. Regulation of 26 S RNA Synthesis

Temperature-sensitive mutants deficient in 26 S RNA synthesis have been isolated from both the Sindbis and SF viruses by mutagen treatment (*Scheele* and *Pfefferkorn*, 1969b; *Atkins* et al., 1974; *Keränen* and *Kääriäinen*, 1974; *Bracha* et al., 1976; *Saraste* et al., 1977). One SF virus mutant, ts-4, which is phenotypically RNA negative, has a reversible temperature-dependent defect in the synthesis of 26 S RNA (*Saraste* et al., 1977). When ts-4 is grown at the permissive temperature (28° C), both 42 S and 26 S RNA are synthesized, but upon shift to restrictive temperature (39° C), only the synthesis of 26 S RNA stops. When the cultures are shifted back to 28° C, the synthesis of 26 S RNA starts again even in the presence of cycloheximide.

The fate of 42 S RNA-negative strands in RF I, II, and III before and after temperature shifts was followed by exposing the cells to ^{3}H-uridine early during infection at 28° C. Further labeling of viral RNAs was stopped using glucosamine and an excess of unlabeled uridine (*Sawicki* et al., 1978). When the 26 S RNA synthesis stopped at 39° C, there was a quantitative shift of label from RF II plus RF III to RF I. When shifted back to 28° C, part of the label from RF I was again found in RF II and RF III (*Sawicki* et al., 1978). These results confirm the proposed interconvertibility between RIa and RIb (*Simmons* and *Strauss*, 1972b), i.e., that 42 S RNA-negative strands used as templates for 26 S RNA synthesis are shifted to templates of 42 S RNA-positive strand synthesis and vice versa.

The interconversion is regulated by a virus-specific protein which is mutated in ts-4. This "interconversion protein" (*Scheele* and *Pfefferkorn*, 1969b) or "26 S protein" (*Keränen* and *Kääriäinen*, 1974) must be different from the RNA polymerase synthesizing 42 S RNA since the synthesis of this RNA is unaffected by the upward shift in ts-4. Another mutant deficient in 26 S RNA has been isolated from the SF virus. This ts-1 mutant has an RNA-positive phenotype and synthesizes the same amount of RNA as the wild type (*Keränen* and *Kääriäinen*, 1974; *Kääriäinen* et al., 1975b). This means that the function of the interconversion protein is neither necessary in the synthesis of 42 S RNA-negative nor 42 S RNA-positive strands. Some RNA-negative ts-mutants of the SF virus can at least partly restore the 26 S RNA deficiency of ts-1 at 39° C, giving further evidence for the different functions of RNA polymerase and interconversion protein (*Keränen*, 1977a). The possibility that the latter is in fact a different polymerase specific for the synthesis of 26 S RNA cannot be excluded at present, but we consider it to be unlikely. Our view of the regulation of 26 S RNA synthesis is presented in Figure 5a. When interconversion protein binds to RIa, possibly to the negative strand template, it creates a single-stranded area or loop, which can be easily hydrolyzed by ribonuclease leading to formation of RF II and RF III (*Simmons* and *Strauss*, 1972b). Those polymerase molecules which had started transcription from the 3′ end of the 42 S RNA-negative strand cannot continue due to alterations in the template caused by the interconversion protein. An internal initiation by new RNA polymerase molecules becomes possible leading to the synthesis of 26 S RNA. Detachment of the factor (e.g., by an upward shift of temperature of ts-4 infected cultures) converts the RIb to RIa allowing the "jammed" polymerase molecules to complete the nascent 42 S RNA-positive strands, which were started before the interconversion protein was attached. The temperature dependence of 26 S RNA synthesis (*Scheele* and *Pfefferkorn*, 1969 b; *Atkins* et al., 1974; *Keränen* and *Kääriäi-nen*, 1974; *Wengler* and *Wengler*, 1976 b; *Sawicki* et al., 1978) would be easily explained if the interconversion protein bound more easily to the template at the lower temperature. At lower temperatures, the binding is favored and more 26 S RNA is produced. The difference in the 42 S/26 S RNA ratios between the Sindbis and SF viruses would also reflect the different binding constants of the interconversion protein of these two viruses. This model would also explain the slow labeling of RF II which would reflect the rate of interconversion between RIa and RIb.

If puromycin is added to the Sindbis virus-infected cultures 90 min after infection, practically no 26 S RNA is formed, whereas the synthesis of 42 S RNA is found in almost normal quantities (*Scheele* and *Pfefferkorn*, 1969 b). Assuming that the interconversion protein is a soluble protein, which in its free form is not associated with membranes or with the replication complex, the result of *Scheele* and *Pfefferkorn* can be understood. Higher concentrations of this protein would be required to initiate the 26 S RNA synthesis than to start the actual RNA polymerase function, which is tightly membrane-bound and most probably also template associated (*Friedman* et al., 1972; *Clewley* and *Kennedy*, 1976). The interconversion protein has not been identified. The recent results by *Bracha* et al. (1976) and ourselves (*Kääriäinen* et al., 1978) suggest that at least one of the nonstructural proteins is involved in this function.

The circular model of regulation of 26 S RNA synthesis shown in Figure 5b cannot be excluded at present. It postulates that the 5′ ends of both 42 S and 26 S RNA would be transcribed from the same sequence in the 3′ end of the 42 S RNA-negative strand. The function of the interconversion protein would be to align the sequences from the 3′ and 5′ ends to create a large loop equivalent to RF II. Transcription beginning from the 3′ end of the negative strand either continues through the loop or "jumps over" the loop. In the former case, newly synthesized positive strands should be specifically destroyed as has been suggested to occur during the processing of adenovirus mRNA (*Berget* et al., 1977; *Chow* et al., 1977; *Gelinas* and *Roberts*, 1977; *Klessig*, 1977). The finding of RF II and RF III in equimolar ratios after long labeling with ^{3}H-uridine speaks against this possibility. The model in which the polymerase molecule jumps over the loop (Fig. 5b) would probably cause the formation of a gap in the synthesized RNA strand and would thus require ligase activity. The validity of the circular model proposed above can be tested by sequencing the 5′ ends of both 42 S and 26 S RNA. If the sequences are different the model is incorrect. Translation of 42 S RNA under some conditions also yields structural proteins (*Smith* et al., 1974; *Glanville* et al., 1976 a), indicating that the initiation site for structural proteins exists in that part of 42 *S* RNA which is equivalent to 26 *S* RNA. These results would speak against the circular model. This model would, however, easily explain the creation of defective interfering RNAs of different lengths, which consist of part of the 5′ terminus and larger or smaller part of the poly (A)-containing 3′ end (*Bruton* and *Kennedy*, 1976; *Bruton* et al., 1976; *Kennedy*, 1976).

4. RNA Polymerase in α-Virus-Infected Cells

The ability to incorporate labeled nucleotides or phosphate into RNA in the presence of actinomycin D can be considered to be a rough measure of RNA polymerase activity in the infected cells. Kinetic studies carried out throughout the infectious cycle have revealed that the viral RNA synthesis is hardly detectable during the first hour postinfection and thereafter slowly increases during the next hour. An exponential increase follows which lasts to 4–5 h postinfection, whereafter the rate of RNA synthesis remains almost constant (*Sonnabend* et al., 1967; *Kääriäinen* and *Gomatos*, 1969; *Simmons* and *Strauss*, 1972 b; *Bruton* and *Kennedy*, 1975; *Wengler* and *Wengler*, 1975 b). Addition of protein synthesis inhibitors after 3 h postinfection affects the RNA synthesizing capacity of the cells little if at all, indicating that enough stable RNA polymerase has been formed (*Friedman* and *Grimley*, 1969; *Scheele* and *Pfefferkorn*, 1969 b; *Ranki* and *Kääriäinen*, 1970; *Wengler* and *Wengler*, 1975 b; *Kääriäinen* et al., 1978) to ensure the continuation of RNA synthesis. The RNA polymerase activity has been demonstrated directly by several investigators in subcellular fractions derived from α-virus-infected cells (*Martin* and *Sonnabend*, 1967; *Martin*, 1969; *Sreevalsan* and *Yin*, 1969; *Ranki* and *Kääriäinen*, 1970; *Sreevalsan*, 1970; *Ershov* et al., 1971; *Takehara*, 1971; *Friedman* et al., 1972; *Michel* and *Gomatos*, 1973).

In all cases, the polymerase-template complex is bound to large-sized intracellular membranes often capable of synthesizing 42 *S* and 26 *S* RNA, replicative intermediates, and double-stranded RNA. Fractionation of the membranes further revealed the presence of typical cytoplasmic vacuoles (CPV I) (*Grimley* et al., 1968) in fractions enriched in RNA polymerase activity (*Friedman* et al., 1972).

Only recently, *Clewley* and *Kennedy* (1976) purified the RNA polymerase activity further from discontinuous sucrose gradient fractions. The solubilized enzyme preparation was applied to an affinity chromatography column in which 42 *S* RNA was bound to oligo(dT)cellulose. By this means, two virus-specific and one host cell protein with apparent molecular weights of 90,000, 63,000, and 40,000 were eluted in a fraction which still stimulated incorporation of ^{3}H-GTP into acid-insoluble form. There was a drastic decrease in enzymatic activity when the membrane fraction was solubilized with Triton X-101, indicating the importance of membranes in protecting the enzyme-template complex against cellular nucleases.

The problems of α-virus-directed RNA synthesis are far from being solved. The function of all four nonstructural proteins has yet to be established either in the synthesis of 42 *S* RNA-negative and positive strands or 26 *S* RNA. The possible contribution of host components in the RNA polymerase(s) is also unknown. The role of viral structural proteins in the regulation of RNA synthesis cannot be excluded either, as suggested by a recent report of *Keränen* (1977b). UV-irradiated ts-mutant of the SF virus was able to interfere with the RNA replication of the wild type, showing that structural proteins, possibly the capsid protein, might directly inhibit the RNA polymerase action.

D. α-Virus-Directed Protein Synthesis

1. Translation of Structural Proteins

The mode of translation of α-virus structural proteins has been clarified to a large extent during the recent past (for review, see *Strauss* and *Strauss*, 1977). The messenger for the structural proteins is the intracellular 26 *S* RNA as has been shown by in vitro translation in different cell-free systems derived from mammalian and wheat germ cells (*Cancedda* et al., 1974a, b; *Clegg* and *Kennedy,* 1974b; *Simmons* and *Strauss* 1974b; *Wengler* et al., 1974; *Clegg* and *Kennedy,* 1975a, b, c; *Glanville* et al., 1976a) (Fig. 6).

Association of 26 *S* RNA with polysomes in infected cells confirms the messenger role of this RNA (*Kennedy, 1972; Mowshowitz,* 1973; *Poiree* et al., 1973; *Söderlund* et al., 1973; *Simmons* and *Strauss,* 1974a; *Wengler* and *Wengler,* 1974; *Wengler* et al., 1974; *Eaton* and *Regnery,* 1975; *Martire* et al., 1977). The 26 *S* RNA polysomes are membrane bound (*Kennedy, 1972; Wirth* et al., 1977) as is the synthesis of structural proteins (*Friedman,* 1968a). The structural proteins are principally translated as a polyprotein with a molecular weight of 130,000–140,000 as shown by three independent lines of evidence:
1. When 26 *S* RNA is translated in vitro in the presence of formyl ^{35}S methionyl-transfer RNA_f, only one labeled tryptic (or pronase) peptide is detected (*Cancedda* et al., 1975; *Clegg* and *Kennedy,* 1975b; *Glanville* et al., 1976b), which yields f-met-asn as a dipeptide (*Glanville* et al., 1976b). The same dipeptide (met-asn) is obtained in the 26 *S* RNA-directed in vitro system when elongation is inhibited by diphtheria toxin (*Clegg* and *Kennedy,* 1975b). Polyacrylamide gel analysis of the product labeled with formyl-^{35}S-methionine in an in vitro

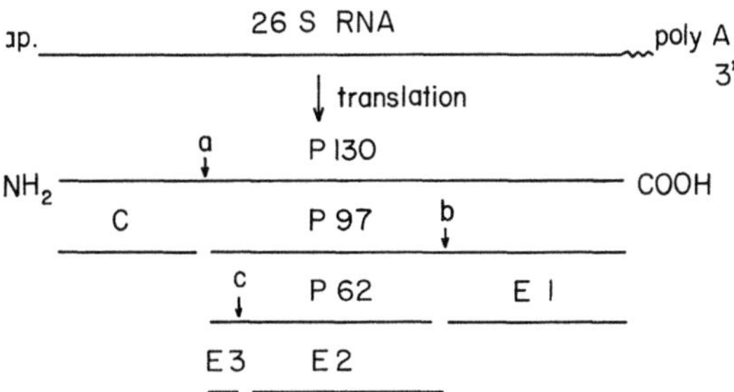

Fig. 6. Translation of structural protein as a polyprotein from the 26 *S* RNA. The cleavage of capsid protein is nascent (a), cleavage between p62 and E1 takes place rapidly after translation (b), whereas p62 is cleaved during the maturation of the virus (c)

system programmed with 26 *S* RNA yields only labeled capsid protein, indicating that this protein is N-terminal in the structural polyprotein.

2. Temperature-sensitive, RNA-positive mutants have been isolated from the Sindbis and SF viruses, which produce a large 130,000–140,000 daltons protein. This protein has the tryptic peptides of the capsid and all envelope proteins (*Schlesinger M.* and *Schlesinger S.*, 1973; *Keränen* and *Kääriäinen*, 1975; *Lachmi* et al., 1975).

3. Sequential labeling of structural proteins takes place when infected cells are released from the hypertonic block of initiation (*Saborio* et al., 1974); capsid protein is labeled first, followed by envelope proteins E3, E2, and finally by E1 (*Clegg,* 1975; *Clegg* and *Kennedy,* 1975 b, c; *Lachmi* and *Kääriäinen,* 1976; *Garoff* and *Söderlund,* 1978).

The gene order of the structural proteins in 26 *S* RNA revealed by sequential labeling has been independently proved by tryptic peptide mapping of some aberrative cleavage products found in cells infected with some temperature-sensitive mutants of the SF virus (*Kääriäinen* et al., 1975 b; *Lachmi* et al., 1975). The capsid protein is cleaved from the nascent polyprotein both in vivo (*Burke,* 1975; *Strauss* and *Strauss,* 1976) and in vitro (*Cancedda* et al., 1974 a, b; *Clegg* and *Kennedy,* 1974 b; *Wengler* et al., 1974; *Glanville* and *Ulmanen,* 1976), probably immediately after the ribosome has finished the translation of the sequence coding for capsid protein (*Clegg* 1975; *Söderlund,* 1976). The nature of the cleavage enzyme is not clear, but recent evidence by *Scupham* et al. (1977) obtained in mixed infections with Sindbis ts-mutants suggests that the enzyme is virus specific. Since the N-terminal amino acid of capsid protein is probably lysine (*Kennedy* and *Burke,* 1972), there must be another cleavage whereby at least met-asn or a longer oligopeptide is removed from the N-terminus of the primary translational product of the capsid protein. This type of "lead-in" sequence was previously demonstrated in the mengovirus-directed polyprotein (*Smith,* 1973).

The cleavage between the envelope proteins takes place at the time when most of the polyprotein has been translated (*Söderlund,* 1976). The products of this cleavage are the envelope protein E1 and the 62,000–68,000 daltons precursor protein of E2 and E3, i.e., the p62 (NSP 68, NVP 68, or pE$_2$) (*Schlesinger S.* and *Schlesinger M.,* 1972; *Simons* et al., 1973 b; *Garoff* et al., 1974; *Lachmi* et al., 1975). The cleavage between the capsid protein and the envelope protein is probably the only one which takes place in reticulocyte lysate programmed with the Sindbis virus 26 *S* RNA (*Simmons* and *Strauss,* 1974 b), leading to production of 100,000 daltons envelope protein "precursor" and capsid protein in almost a 1:1 molar ratio. If membranes from isolated endoplasmic reticulum are added to the incubation mixture, cleavage between E1 and p62 seems to take place (*Dobberstein,* 1977), showing the importance of membranes in the processing of these proteins.

The capsid protein attaches to the 60 *S* ribosomal subunit in the polysomes immediately after its synthesis (*Glanville* and *Ulmanen,* 1976; *Ulmanen* et al., 1976), while the p62 (pE$_2$) and E1 associate with membranes of the endoplasmic reticulum and become glycosylated, as will be discussed later.

The possibility of premature termination after the translation of capsid protein has been suggested as an explanation for excessive synthesis of capsid protein in Sindbis virus-infected cells (*Cancedda* and *Schlesinger*, 1974). This may not be a general rule in α-virus protein synthesis since in SF virus-infected cells, capsid and envelope proteins are synthesized in about a 1:1 molar ratio (*Morser* et al., 1973; *Morser* and *Burke*, 1974; *Keränen* and *Kääriäinen*, 1975). The overproduction of capsid protein may also not be the rule in the Sindbis virus-infected cells, as judged from the earlier published results by several investigators (*Strauss* et al., 1969; *Scheele* and *Pfefferkorn*, 1970; *Pfefferkorn* and *Boyle*, 1972; *Snyder* and *Sreevalsan*, 1974).

The molecular weight of 26 S RNA has been estimated to be between 1.6 and 1.8×10^6 (*Levin* and *Friedman*, 1971; *Simmons* and *Strauss*, 1972a; *Martin* and *Burke*, 1974). Its theoretic coding capacity is thus at least 160,000 daltons of protein. From this capacity, about 130,000 is used to code for the known structural proteins. It will be interesting to see whether hithertho unknown proteins can be demonstrated in the α-virus-infected cells in the future.

2. Translation of Nonstructural Proteins

Recently, several other-than-structural proteins have been detected in SF and Sindbis virus-infected cells (*Lachmi* et al., 1975; *Bracha* et al., 1976; *Kaluza*, 1976; *Kaluza* et al., 1976; *Brzeski* and *Kennedy*, 1977). These proteins are designated as nonstructural proteins until their function has been definitely established. Two nonstructural proteins with apparent molecular weights of 86,000 and 72,000 were first identified by tryptic peptide mapping from cells infected with a temperature-sensitive mutant, ts-1, of the SF virus (*Lachmi* et al., 1975). In ts-1-infected cells, these proteins are produced in an excessive amount compared to the wild type (*Keränen* and *Kääriäinen*, 1975; *Kääriäinen* et al., 1975a, b).

When ts-1-infected cell cultures were released from the hypertonic block of initiation, sequential labeling of four stable nonstructural proteins with molecular weights of 70,000 (ns70), 86,000 (ns86), 72,000 (ns72, previously ns78), and 60,000 (ns60) were demonstrated (*Lachmi* and *Kääriäinen*, 1976). Two short-lived large proteins with molecular weights of about 155,000 (ns155 or A) and 135,000 (ns135 or B) were also labeled. The former was seen after a short labeling, whereas ns135 appeared only after 10 min or a longer labeling period, following synchronous initiation of protein synthesis. These results suggested a precursor-product relationship between the large short-lived proteins and the more stable ones. The evidence for the existence of ns60 is still indirect and based on the labeling of a protein of this size in pactamycin-treated cells in which the structural protein synthesis was already stopped (*Lachmi* and *Kääriäinen*, 1976). Comparison of tryptic peptides of ns155, ns135, ns70, ns86, and ns72 has to a large extent confirmed the proposed precursor-product relationship. First, the primary structures of ns155 and ns135 are different, indicating that the total molecular weight of the nonstructural proteins is close to 300,000 (*Glanville* and *Lachmi*, 1977). Secondly, the tryptic peptides of ns70 and ns86 are found in ns155 and those of ns72 are found in ns135 (*Glanville* et al., 1978). In ns135, there are peptides not found in ns72, and these may be derived from ns60.

The sequential labeling of the different nonstructural proteins suggested that they are translated as a polyprotein in analogy to the structural proteins (*Lachmi* and *Kääriäinen, 1976*). Support of this view has been obtained from the results of translation of 42 *S* RNA in vitro in the presence of formyl ^{35}S-methionyl-transfer RNA_f. Analysis of tryptic (or pronase) peptides of the translational product yielded only one labeled peptide, suggesting that the translation of 42 *S* RNA is initiated from a single initiation site (*Glanville* et al., 1976b). The initiation dipeptide was identified to be f-met-ala. Similar results have been reported by *Cancedda* et al. (1975), although they found small amounts of radioactivity which evidently was derived from an internal initation site, respective to the initiation site of 26 *S* RNA.

The viral origin of the nonstructural proteins of SF virus, at least of ns155 (i.e., ns70 and ns86) and partly of ns72, has been confirmed by comparing the tryptic peptides of the above proteins with those derived from the translational product of 42 *S* RNA in wheat germ cell-free extract (*Glanville* and *Lachmi*, 1977).

The presence of the nonstructural proteins has also been demonstrated in the SF virus wild type-infected cells in which their maximum rate of synthesis takes place between 3 and 4 h postinfection, declining thereafter (*Lachmi* and *Kääriäinen*, 1977). *Clegg* et al. (1976) have shown two nonstructural proteins with molecular weights of 90,000 and 63,000. These proteins are probably equivalent to ns86 and ns70. They are presumably formed from larger precursors which have molecular weights of 200,000, 184,000, and 150,000.

The 42 *S* RNA is most probably the messenger for the nonstructural proteins in the infected cells, since this RNA has been found in the polysomes by several investigators (*Mowshowitz, 1973; Söderlund* et al., 1973; *Simmons* and *Strauss, 1974a; Wengler* and *Wengler, 1975a; Martire* et al., 1977). *Bracha* et al. (1976) have shown the existence of a 200,000 daltons nonstructural protein in cells infected with the temperature-sensitive mutants ts-21 and ts-24 of the Sindbis virus. The protein accumulated only at the restrictive temperature, suggesting a cleavage defect of the nonstructural polyprotein with these RNA negative mutants.

Brzeski and *Kennedy* (1977) have found several nonstructural proteins with molecular weights of 230,000, 215,000, 150,000, 89,000, 82,000, 76,000, and 60,000 in Sindbis virus wild type-infected cells. The large proteins accumulated when zinc ions were used to inhibit the cleavages of the nonstructural polyprotein (*Bracha* and *Schlesinger, 1976a*). The ns89, ns82, and ns60 are stable products, whereas the others are their precursors. Exposure of the infected cells to ^{35}S-methionine after release from the hypertonic block of initiation labeled the proteins sequentially in the order ns60, ns89, and ns82, suggesting that they are made from a large polyprotein which the authors think is ns230.

A 220,000 daltons protein (ns220) was recently demonstrated with a temperature-sensitive mutant, ts-4, of the SF virus (*Kääriäinen* et al., 1978). This protein is detected only at the restrictive temperature as is the case with the similar size protein from the Sindbis virus ts-mutants (*Bracha* et al., 1976). Short pulses given after release from the hypertonic block of initiation labels the ns220 as well as ns155, suggesting that both are derived from the N-terminal part of the nonstructural polyprotein.

Waite (1973) has reported an interesting result using an RNA-negative mutant, ts-11, of the Sindbis virus. When ts-11-infected cultures were shifted to the restrictive temperature, a protein with an apparent molecular weight of 133,000 accumulated into the cells. This protein remained stable even if the cultures were shifted back to the permissive temperature. *Waite* suggested that the large protein was the structural polyprotein. It would be more plausible to assume that the 133,000 daltons protein is a nonstructural precursor protein, perhaps similar to that of ns135 of the SF virus, which accumulates because of a cleavage defect caused by a mutation in one of the nonstructural proteins.

Kaluza has demonstrated the presence of at least some of the nonstructural proteins in SF virus-infected cells pretreated with fowl plague virus which shuts off the host protein synthesis (*Kaluza,* 1976; *Kaluza* et al., 1976). *Friedman* (1968c) may have detected some of the nonstructural proteins in quanidine-treated SF virus-infected cells. The identification of the proteins is, however, difficult because of the different polyacrylamide gel systems used.

The mode of translation of α-virus nonstructural proteins is nearly solved. In spite of the slightly different results obtained with SF and Sindbis viruses, there are also striking similarities:

1. Results with both viruses suggest that the nonstructural proteins are translated as a large polyprotein of at least 230,000 but probably 290,000 molecular weight.
2. The identification of the different stable nonstructural proteins is facilitated by their order of labeling after synchronous initiation. Thus the N-terminal protein, ns70, in the SF virus has its counterpart in ns60 in Sindbis virus-infected cells, followed by ns86 (SF virus), ns89 (Sindbis virus) and ns72 (SF virus), ns82 (Sindbis virus) (Fig. 7a, b).

The existence of the fourth nonstructural protein has been difficult to show in either virus, due to its migration at the position of p62 (pE$_2$). The differences in apparent molecular weights obtained by research groups may be real, but they may also reflect technical differences. For example, with the SF virus, ns72, ns70, and p62 migrate together in the discontinuous polyacrylamide gel system by *Laemmli* (1970), whereas ns72 migrates far from p62 and allows the detection of ns70 between ns72 and p62 in the *Neville* gel system (1971) (unpublished results). The different primary structures of ns155 and ns135 strongly suggest that a fourth nonstructural protein exists, and thus we should be able to find the whole nonstructural polyprotein in the future.

As mentioned previously, two of the nonstructural proteins, ns90 (probably identical with our ns86 and the ns89 of the Sindbis virus) and ns63 (possibly our ns70 and equivalent to ns60 of the Sindbis virus), are associated with purified preparations of RNA polymerizing activity (*Clegg* et al., 1976; *Clewley* and *Kennedy,* 1976). The functions of the other nonstructural proteins remain to be solved in the future (*Pfefferkorn,* 1975).

3. Control of Protein Synthesis in α-Virus-Infected Cells

A drastic inhibition of host cell protein synthesis takes place in vertebrate cells infected with different α-viruses (*Lust,* 1966; *Hay* et al., 1968; *Strauss* et al., 1969; *Igarashi,* 1970; *Mussgay* et al., 1970; *Takehara,* 1972; *Keränen* and *Kääriäinen,* 1975; *Lubiniecki,* 1975), usually between 3–5 h postinfection.

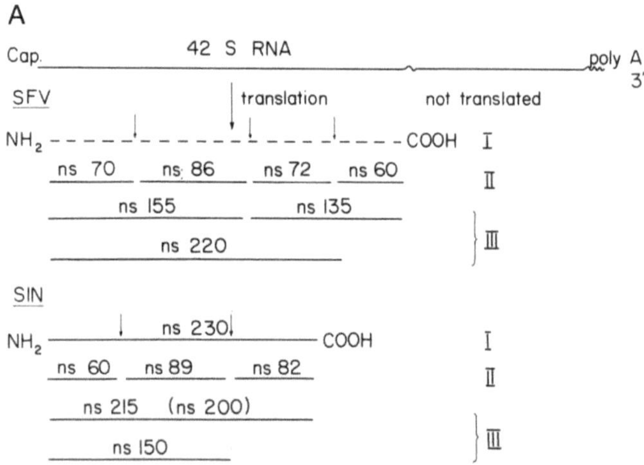

Fig. 7A. Translation of Semliki Forest (SFV) and Sindbis virus (SIN) nonstructural proteins from 42 *S* RNA. I indicates the postulated primary translational product, II the stabile cleavage products, and III the detected intermediates. Arrows indicate the cleavage sites. Dotted line refers to a protein which has not been found

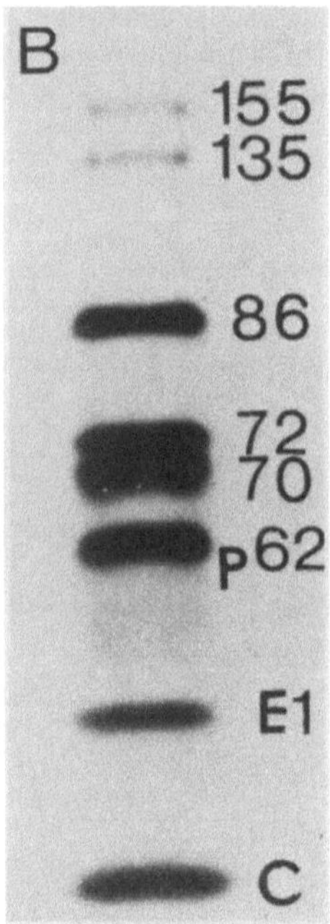

Fig. 7B. Fluorogram of polyacrylamide gel showing SF virus-specific structural and nonstructural proteins

The mechanism of this inhibition is not known. Competition between viral and host messenger RNAs has been suggested to explain the host cell inhibition (*Keränen* and *Kääriäinen,* 1975; *Wengler* and *Wengler,* 1976b; *Atkins,* 1976). *Tuomi* et al. (1975) calculated that there are about equal amounts of viral and cellular mRNAs competing for the ribosomes at 4 h postinfection. This indicates

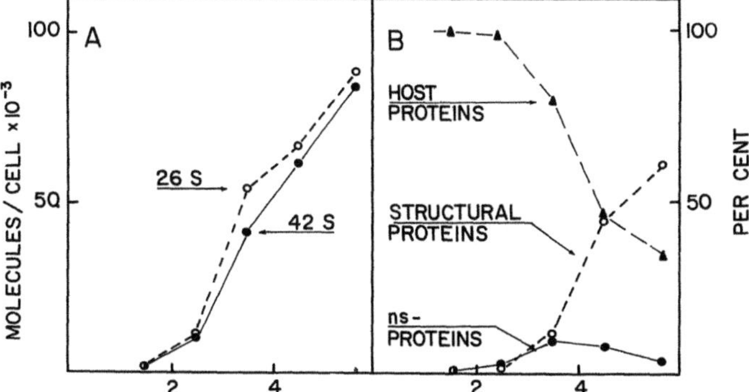

Fig. 8. RNA and protein synthesis in SF virus-infected cells according to *Tuomi* et al. (1975) and *Lachmi* and *Kääriäinen* (1977). (A) The number of virus-specific RNAs is compared to the rate (B) of synthesis of host- and virus-specific proteins

that the affinity of ribosomes to the viral messenger is greater than to the host cell messenger. The required difference in affinity may be so small that it would be difficult to verify experimentally in vitro, as has turned out to be the case in cell-free extracts from picornavirus-infected cells (*Lawrence* and *Thach*, 1974; *Abreu* and *Lucas-Lenard*, 1976).

Another type of protein synthesis control in α-virus-infected cells has been described (*Kääriäinen* et al., 1976; *Lachmi* and *Kääriäinen*, 1977). The maximum rate of synthesis of nonstructural proteins is reached between 3 and 4 h postinfection at a time when only about 50% of 42 *S* RNA has been synthesized. Thereafter, the rate of synthesis of the nonstructural proteins declines continuously despite the increase in the amount of 42 *S* RNA. The efficiency of the 42 *S* RNA as a messenger declines by an overall factor of six. This decline is at least partly due to an increased encapsidation rate of 42 *S* RNA molecules into nucleocapsids (*Ulmanen* et al., unpublished results). However, a nucleocapsid-negative, RNA-positive mutant of the SF virus (ts-3) shows the same shutoff (*Keränen* and *Kääriäinen*, 1975) in the synthesis of nonstructural proteins as does the wild type, indicating that mechanisms other than encapsidation of 42 *S* RNA must be involved in this translational control. The synthesis of structural proteins does not seem to be regulated at the translational level, since their rate of synthesis closely reflects the amount of 26 *S* RNA in the infected cells (*Lachmi* and *Kääriäinen*, 1977) (Fig. 8).

The excessive synthesis of nonstructural proteins in the SF virus ts-1 mutant-infected cells has been studied to some extent (*Kääriäinen* et al., 1976). Ts-1-infected cells initially maintained at the permissive temperature accelerate the synthesis of nonstructural proteins only slowly when shifted to the restrictive temperature. If the infection is started at 39° C and the cultures thereafter shifted to 28° C, the shut off of nonstructural protein synthesis takes about 2 h. The synthesis continues somewhat longer if cycloheximide is added at the moment of shift-down. These observations would suggest that the factors controlling the shut off must be synthesized after the shift-down and that a fairly large amount of these factors must be made before most of the 42 *S* RNAs are shut off from the synthesis of nonstructural proteins. The shift-up experiment suggests that those 42 S RNA molecules which had been shut off from the synthesis of nonstructural proteins at 28° C are not able to start it again when shifted to 39° C. Rather, it would mean that new 42 S RNA molecules are needed which are not affected by the control mechanism operating at the permissive temperature. These experiments rule out a temperature-sensitive reversible control mechanism similar to that seen in the regulation of 26 S RNA synthesis discussed earlier.

E. Assembly of Nucleocapsid

The assembly of α-viruses differ from most icosahedral plant and animal viruses because no empty nucleocapsids can be seen in the infected cells (*Acheson* and *Tamm,* 1967; *Erlandson* et al., 1967; *McGee-Russell* and *Gosztonyi,* 1967) nor have fractionation of infected cells revealed RNA-free structures or "top components" which could be regarded as empty capsids (*Ben-Ishai* et al., 1968; *Friedman,* 1968a; *Friedman* and *Grimley,* 1969; *Söderlund,* 1973).

Kinetic studies carried out by amino acid labeling have shown that the capsid protein is transferred to the cytoplasmic nucleocapsids within 2–10 min (*Söderlund,* 1973) and to released virions within 10–30 min (*Scheele* and *Pfefferkorn,* 1969a). Free capsid protein cannot be demonstrated even after short pulses, suggesting that capsid protein binds rapidly to sedimenting structures. After short pulses with ^{35}S-methionine, most of the newly formed capsid protein was associated with polysomes from which it could be chased to nucleocapsids. This process was inhibited if cycloheximide was added after the pulse (*Söderlund,* 1973). The inhibition of nucleocapsid assembly by cycloheximide was first demonstrated by *Friedman* and *Grimley* (1969).

The recent findings that nascent capsid protein binds to the large ribosomal subunit in the polysomes explains why no free capsid protein has been found (*Ulmanen* et al., 1976). Pulse-chase experiments show that the capsid protein is rapidly transferred from the large subunits to 42 S RNA (*Söderlund* and *Ulmanen,* 1977). The transfer takes place with two different kinetics; initially about 30% of the pulse-labeled capsid protein is chased within 2 min, whereafter the transfer is considerably slower. This result must mean that the initial transfer from the large ribosomal subunit to 42 S RNA takes place at the polysomes. The second phase may take place after the ribosomes have been released into the monosome pool. How the transfer of capsid protein takes place is not known. The kinetics suggest that the nucleocapsid may be assembled by the continuous addition of capsid protein to 42 S RNA, but no intermediates or nucleocapsid precursors have been isolated so far. The possible linkage between polysomes and assembling nucleocapsids could explain why nucleocapsids have been isolated from the polysomal fraction (*Mowshowitz,* 1973; *Simmons* and *Strauss,* 1974a).

The attachment of capsid protein is specific for 42 S RNA since 26 S RNA does not bind this protein (*Ulmanen* et al., 1976) in contrast to predictions based on earlier experiments (*Söderlund* and *Kääriäinen,* 1974). This means that the 5' two-thirds of the 42 S RNA, which is different from the 26 S RNA, has a specific binding region for capsid protein. These sequences must have been preserved in the defective interfering RNAs because these are encapsidated (*Bruton* and *Kennedy*, 1976).

The information available on the assembly of α-virus nucleocapsids suggests that it is a unique process. It will be interesting to see whether similar assembly processes will be found among those small isometric plant viruses, like cucumber chlorotic mottle viruses, which apparently have strong RNA-protein interactions as the stabilizing force in the virion (*Kaper*, 1975).

Nucleocapsid seems to be formed in great excess compared to the amount which actually is released from the cell as mature virus. Intermediate and late in the infection, most of the genome RNA is encapsidated (*Ulmanen*, personal communication), but only 5%–10% is released (*Tuomi* et al., 1975). Nucleocapsids accumulate to such an extent that paracrystalline arrays can be visualized by electron microscopy in the cytoplasm of infected cells (*Morgan* et al., 1961; *Acheson* and *Tamm*, 1967; *Wagner* et al., 1975).

F. Assembly of Viral Envelope

In this section, we will mainly discuss the most recent findings since the earlier literature was recently reviewed by *Kääriäinen* and *Renkonen* (1977) and *Simons* et al. (1978).

1. Glycosylation of Envelope Proteins

The structural proteins are translated from 26 S RNA in membrane-bound polysomes (*Wirth* et al., 1977). The envelope proteins are inserted into the membranes of endoplasmic reticulum and protrude through the membrane into the cysternal side of the rough endoplasmic reticulum (Fig. 9). The nascent chains of p62 (pE_2) and E1 are glycosylated by lipid-containing oligosaccharide intermediates, probably a dolichol oligosaccharide (*Krag* and *Robbins,* 1977; *Sefton*, 1977). In this process, the glucosamine-containing mannose-rich core of the A-oligosaccharide type chain (*Johnston* and *Clamp*, 1971; *Spiro*, 1973) and the probably identical B-type chain are transferred to the envelope proteins. If primary glycosylation is inhibited by tunicamycin, which is a specific inhibitor of oligosaccharide lipid intermediate synthesis (*Takatzuki* et al., 1975; *Tkacz* and *Lampen*, 1975; *Struck* and *Lennarz*, 1977), envelope proteins p62 (pE_2) and E1, which have increased mobility in polyacrylamide gels, are found from infected cells (*Schwarz* et al., 1976; *Leavitt* et al., 1977). The same "apoproteins" are formed if the cells are incubated in glucose-free medium (*Kaluza,* 1975; *Sefton*, 1977). These may not be able to accept oligosaccharides, e.g., due to a wrong configuration (*Sefton*, 1977). Similar apoproteins, or poorly glycosylated envelope proteins, are synthesized in the presence of glucosamine (*Duda* and *Schlesinger*, 1975), 2-deoxy-D-glucose (*Kaluza* et al., 1973; *Scholtissek* and *Kaluza*,

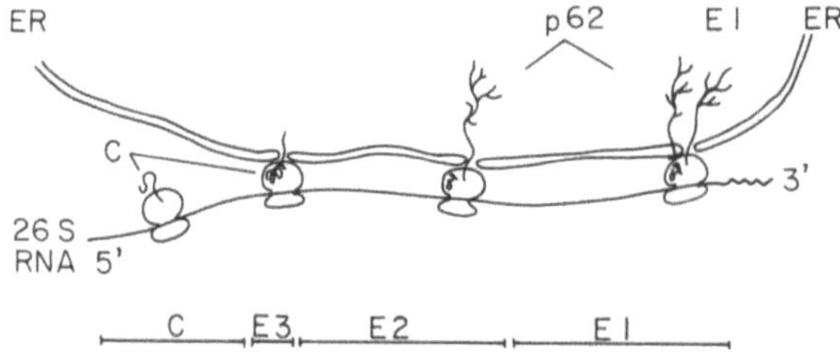

Fig. 9. The structural proteins of α-viruses are translated from 26 *S* RNA on membrane-bound polysomes on the cytoplasmic side of the endoplasmic reticulum membranes (ER). Translation of capsid protein takes place on free ribosomes. After completion of translation of the capsid protein, nascent cleavage takes place and the protein attaches to the 60*S* ribosomal subunit. The nascent N-terminus of the envelope protein precursor directs the binding of the ribosome to the ER membrane, and the growing polypeptide chain is protruded through the membrane to the cisternal side and becomes glycosylated by lipid-bound oligosaccharide intermediates. Cleavage between p62 (pE$_2$) and E1 takes place but the two proteins remain closely associated. The envelope glycoproteins remain in the membrane by their C-terminal hydrophobic fragments and can only move at the plane of the membrane

1975; *Scholtissek* et al., 1975; for review see *Scholtissek*, 1975), 2-deoxy-2-fluoro-D-glucose, and mannose (*Schmidt* et al., 1976).

These studies imply that the cleavage between envelope protein E1 and p62 takes place even if glycosylation is inhibited, probably as a result of penetration through the endoplasmic reticulum membrane. As an exception to this rule, the common envelope protein precursor, the 100,000 daltons B protein (*Strauss* et al., 1969; *Schlesinger M.* and *Schlesinger S.*, 1973), which is not glycosylated (*Sefton* and *Burge*, 1973), is protected from the action of added proteases in vitro (*Wirth* et al., 1977), suggesting that it has been transported through the endoplasmic reticulum membrane without cleavage and glycosylation.

The addition of the distal sugars of the A-type chain, such as glucosamine, galactose, and sialic acid, takes place later within approximately 20 min (*Sefton*, 1977). At this stage, part of the mannose residues are probably removed from the "core" before addition of the distal sugars (*Sefton* and *Burge, 1973; Rasilo* and *Renkonen*, 1978). The most exposed glycans like the A-type chain of E3 of the SF virus may be more easily glycosylated and is, therefore, larger than A-type chains of E1 and E2 (*Renkonen*, personal communication). The somewhat incomplete glycosylation does not affect the transport of envelope proteins to the plasma membrane and incorporation into the virions, as has been shown by *Schlesinger* et al. (1976). The Sindbis virus was grown in cells deficient in *N*-acetyl-glucosaminyl transferase activity with normal yields of infectious virus.

2. Transport of Envelope Proteins to the Plasma Membrane

The envelope proteins E1 and E2 have hydrophobic tails by which they are attached to the lipid bilayer in the virus envelope (*Uterman* and *Simons, 1974*). The same interaction must take place in intracellular membranes after the polypeptides have been completed. In both E1 and E2, the hydrophobic regions are close to the C-terminus, as recently shown by *Garoff* and *Söderlund* (1978), making the association to the membranes the final event in the synthesis of polypeptide. Unlike the secretory glycoproteins (*Palade*, 1975), the envelope glycoproteins must migrate in the cells by lateral diffusion at the plane of the membrane (*Hirano* et al., 1972).

Schlesinger and *Schlesinger* (1972) showed by pulse-chase experiments, devised by *Scheele* and *Pfefferkorn* (1969a), that both E1 and E2 were released in the virus particles to the medium with similar kinetics. This suggests that p62 and E1 remain associated with each other after their synthesis and are transported together to the plasma membrane and finally to the virus. Support of this idea was obtained from studies with temperature-sensitive mutants of the Sindbis virus belonging to complementation groups D and E (*Burge* and *Pfefferkorn*, 1966b; 1968). All the mutants show the same defect, namely the inability to cleave p62 (pE$_2$) at the restrictive temperature (*Jones* et al., 1974; *Bracha* and *Schlesinger*, 1976b; *Jones* et al., 1977; *Smith* and *Brown*, 1977; for similar observations with SFV mutants see *Keränen* and *Kääriäinen*, 1975).

In cells infected with ts-20 mutant of the Sindbis virus (with postulated defect in E2 protein) (*Jones* et al., 1974), E1 protein is transported to the outside of the plasma membrane. This is evident since the cells adsorb goose erythrocytes (*Burge* and *Pfefferkorn*, 1968) and E1 can be labeled with lactoperoxidase-mediated iodination as in the wild type infected cells (*Sefton* et al., 1973; *Smith* and *Brown*, 1977). The cells also show membrane fluorescence with antiserum against E1 (*Bell* and *Waite*, 1977). The p62 is present in the plasma membrane fractions (*Jones* et al., 1974), but it can neither be labeled with the lactoperoxidase technique (*Smith* and *Brown*, 1977) nor fluorescent antibodies against E2 (*Bell* and *Waite*, 1977). The crucial demonstration that p62 is on the outside of the plasma membrane would require other methods, e.g., specific antiserum against p62.

The cleavage of p62 (pE$_2$) in wild type-infected cells can be prevented by antiserum against E1, strongly suggesting that both E1 and p62 are on the outside of the plasma membrane and indeed interact so that inhibition of the free mobility of E1 affects p62 so that it cannot be cleaved (*Bracha* and *Schlesinger*, 1976b; *Smith* and *Brown*, 1977). Further support for the interaction between E1 and p62 was obtained from studies with ts-mutants of complementation group D (ts-10 and ts-23), which are supposed to have the defect in the hemagglutinating E1 protein (*Burge* and *Pfefferkorn*, 1968; *Yin*, 1969; *Dalrymple* et al., 1976).

Neither iodination with lactoperoxidase nor ferritin-antibody labeling could demonstrate the envelope proteins on the outside of the plasma membrane (*Smith* and *Brown*, 1977), suggesting that transport to the outside of the plasma membrane of both proteins had failed due to the mutation in E1. This observation also explains why nucleocapsid cannot be seen attached to the cytoplasmic side of the plasma membrane in ts-23-infected cells maintained at the restrictive temperature (*Brown* and *Smith*, 1975). A similar mutant of the WEE virus was reported recently by *Hashimoto* et al. (1977).

Birdwell and *Strauss* (1974a) have nicely demonstrated the presence of envelope proteins at the surface of Sindbis wild type-infected cells as early as 2 h postinfection. The proteins are evenly distributed and patched, easily showing that they are free to move laterally at the plane of the membrane. The amount of envelope proteins on the cell surface increased throughout the infection, but their distribution remained random, showing no prevalent areas of "viral patches" (*Lenard* and *Compans, 1974*).

G. Maturation of α-Virus

Morphogenetic studies with different α-viruses have shown that viral nucleocapsid aligns below the plasma membrane and that the virus is formed by budding through the plasma membrane (*Acheson* and *Tamm*, 1967; *Erlandson* et al., 1967; *Grimley* and *Friedman*, 1970; *Grimley* et al., 1972; *Bykovsky* et al., 1969; *Lascano* et al., 1969; *Tan*, 1970; *Waite* et al., 1972; *Birdwell* et al., 1973; *Gil-Fernandez* et al., 1973; *Virtanen* and *Wartiovaara*, 1974). At this time, there are only projections visible at the site of budding (*Acheson* and *Tamm*, 1967).

The transmembrane interaction between nucleocapsid and the envelope proteins demonstrated in SF virus particles by *Garoff* and *Simons* (1974) would explain how the nucleocapsid can recognize the envelope proteins at the other side of the plasma membrane. The interactions created between nucleocapsid and envelope proteins prevent the free lateral movement of the latter but not the movement of the host cell plasma membrane proteins, which become excluded from the maturing virus particle. If the free lateral mobility is inhibited, e.g., by plant lectins (*Becht* et al., 1971; *Birdwell* and *Strauss*, 1973) or by specific antiserum, virus maturation is inhibited (*Bracha* and *Schlesinger*, 1976b; *Jones* et al., 1977; *Smith* and *Brown*, 1977). The driving force in the assembly is the increasing number of interactions leading to the budding and finally release of the virus particle from the cell membrane (*Brown* et al., 1972; *v. Bonsdorff* and *Harrison*, 1975). During the budding, the virus obtains the lipids from the host cell plasma membrane (*Renkonen* et al., 1971; 1972a, b; 1974; *Laine* et al., 1972; *Hirschberg* and *Robbins*, 1974). *Richardson* and *Vance* (1976) recently demonstrated how labeled proteins from the microsomal fractions are transferred to plasma membrane fraction and finally to released virus particles.

In *Aedes albopictus* cells, the α-viruses probably mature at the intracellular membranes, budding into vaculoles, which release the virus particles into the medium by fusing with the plasma membrane (*Whitfield* et al., 1971; *Raghow* et al., 1973; *Gliedman* et al., 1975; see also *Strauss* and *Strauss*, 1977). The lipids of SF grown in *A. albopictus* cells are drastically different from those derived from BHK 21 cells (*Renkonen* et al., 1974; *Luukkonen* et al., 1976), demonstrating that the host cell specifies the lipid components, whereas the viral genome is responsible for the proteins (*Pfefferkorn* and *Clifford*, 1964; *Luukkonen* et al., 1977b). Variation in the oligosaccharides of the glycoproteins is also a host-specific phenomenon, as has been shown by several investigators (*Keegstra* et al., 1975; *Schlesinger S.* et al., 1976; *Stollar* et al., 1976).

The amino acid sequence of the envelope proteins determines the number and type of the oligosaccharide chains which are added to the glycoproteins by cellular enzymes (*Burge* and *Huang*, 1970; *Grimes* and *Burge*, 1971; *Froger* and *Louisot*, 1972a, b; *Sefton*, 1976). The recognition of envelope proteins through the membrane is sometimes inaccurate, leading to incorporation of several nucleocapsids into giant particles (*Simizu* et al., 1973; *Hashimoto* et al., 1975, 1977; *Strauss E.* et al., 1976, 1977). The reason for this "inaccuracy" may be mutation in the capsid protein rather than envelope proteins, at least in one of the cases studied more closely (*Strauss E.* et al., 1977).

Simultaneous infection of mouse peritoneal macrophages with the Sindbis and lactic dehydrogenase viruses (a nonarbotoga virus) may result in phenotypic mixing, i.e., Sindbis genotypes with LDH envelopes, suggesting that the capsid-envelope protein interaction is not absolutely specific (*Lagwinska* et al., 1975). The phenotypic mixing occurs easily among the α-viruses but cannot be accomplished between α- and rhabdoviruses (*Burge* and *Pfefferkorn*, 1966c; *Choppin* and *Compans*, 1970). The concept of virus maturation presented above is mostly based on experiments performed with cultured cells. The process may be much more complicated in the living organism (*Johnson*, 1965; *Murphy* et al., 1970; *Pathak* and *Webb*, 1974; *Pathak* et al., 1976).

IV. Defective Interfering Particles in α-Virus-Infected Cells

Serial passage of the Sindbis and SF viruses with high multiplicities of infection results in an abrupt drop in infectivity and hemagglutination titers in the culture medium. This autointerference phenomenon is due to accumulation of defective interfering particles (*Schlesinger S.* et al., 1972, 1975; *Eaton* and *Faulkner*, 1973; *Inglot* et al., 1973, 1977; *Shenk* and *Stollar*, 1972, 1973a, b; *Weiss* and *Schlesinger*, 1973; *Eaton*, 1975; *Johnston* et al., 1975; *Bruton* and *Kennedy*, 1976). The progeny particles isolated from the medium after high multiplicity passages interfere with the multiplication of homologous but not with heterologous standard viruses (*Bruton* and *Kennedy*, 1976). The particles with interfering properties (DI particles) have all the proteins of standard viruses in the same proportions but usually contain RNAs which are smaller than the 42 S RNA (*Schlesinger S.* et al., 1974, 1975; *Johnston* et al., 1975; *Bruton* and *Kennedy*, 1976; *Kennedy* et al., 1976; *Guild* and *Stollar*, 1977).

In some cases, the DI particles can be separated from the standard virus by size (*Johnston* et al., 1975) or by density (*Bruton* and *Kennedy*, 1976; *Shenk* and *Stollar*, 1973a). Purified DI particles alone cannot direct any virus-specific functions except adsorption, penetration, and uncoating (*Bruton* et al., 1976), which are probably host functions (*Bracha* et al., 1977). The DI particles isolated so far require the functions of standard viruses to manifest their effects (*Schlesinger S.* et al., 1972; *Schenk* and *Stollar*, 1973a, b; *Schlesinger S.* et al., 1974, 1975; *Bruton* and *Kennedy*, 1976).

A typical effect of DI particles, in addition to their interfering ability, is the stimulation of synthesis of single-stranded RNAs different from the normal

42 *S* and 26 *S* RNA (*Weiss* et al., 1974; *Schlesinger S.* et al., 1974, 1975; *Bruton* et al., 1976). The predominant species has been a single-stranded RNA sedimenting at about 20 *S* (mol wt 0.8–1.0 × 10⁶). Abnormal single-stranded RNA formation is accompanied by a significant reduction in the synthesis of 42 *S* and 26 *S* RNA. The aberrative RNAs are apparently synthesized from different templates than the normal α-virus-specific RNAs, since double-stranded RNAs sedimenting at 12 *S*–16 *S* (*Eaton* and *Faulkner*, 1973; *Guild* and *Stollar*, 1975; *Guild* et al., 1977), probably derived from specific replicative intermediates, have been regularly found (*Bruton* et al., 1976).

The single-stranded DI-RNAs are of positive polarity but cannot stimulate protein synthesis in vitro (*Weiss* et al., 1974; *Bruton* et al., 1976; *Guild* and *Stollar*, 1977). They contain poly (A) and different lengths of sequences from the 3′ end of 42 *S* RNA, which is identical to 26 *S* RNA (*Schlesinger S.* et al., 1974, 1975; *Weiss* et al., 1974; *Bruton* et al., 1976; *Kennedy* et al., 1976; *Guild* and *Stollar*, 1977). The single-stranded DI-RNAs can be derived predominantly from the "26 *S*" RNA region (*Kennedy* et al., 1976) or from the region of 42 *S* RNA, which codes for nonstructural proteins (*Bruton* et al., 1976; *Kennedy* et al., 1976). In both cases, they contain sequences from the 5′ end of 42 *S* RNA (*Kennedy*, 1976; *Kennedy* et al., 1976).

The synthesis of single-stranded DI-RNAs must originally take place from the 42 *S* RNA-positive or negative strand. One possibility is that they are transcribed from the 42 *S* RNA-negative strand by formation of a loop, which the RNA polymerase does not transcribe but rather "jumps over" (Fig. 10). Once formed, the DI-RNA can autoreplicate, using the RNA polymerase provided by the standard virus. The interfering property of the DI-RNAs is considered to be their ability to prevalently bind to the RNA polymerase (*Schlesinger* et al., 1975; *Bruton* et al., 1976; see also *Cole* and *Baltimore*, 1973; *Huang*, 1973). The binding ability and the structure of RNA must be interrelated. It is evident from the work done so far that the 3′ end containing poly (A) cannot be the only signal for the RNA polymerase. If it were so, the 26 *S* RNA should be replicated independently. The finding that part of the 5′ end of 42 *S* RNA is present in the DI-RNAs suggests that both ends are necessary for the binding of polymerase. This conclusion would again speak against the possibility that 26 *S* RNA has an identical 5′ end with the 42 *S* RNA discussed in the context of RNA replication, favoring the model of internal initiation in the synthesis of 26 *S* RNA.

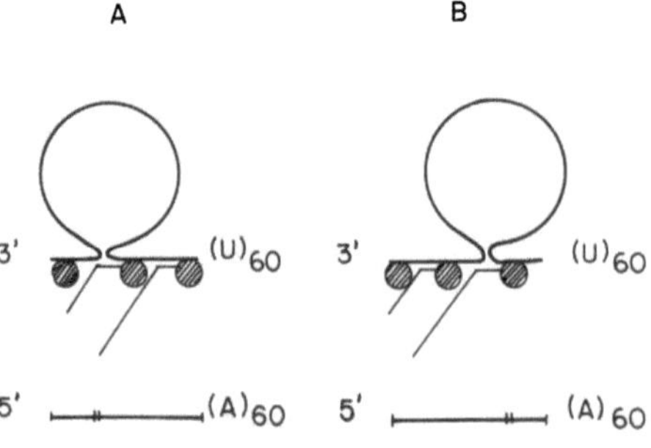

Fig. 10. Hypothetic model for creation of two different types of defective interfering RNAs by transcription of 42 *S* RNA-negative strand template

The presence of the 5′ end of 42 *S* RNA in the DI-RNAs raises the question of the presence of "caps" in these RNAs. Why do they not serve as messenger, even for incomplete products of translation, as the poliovirus DI-RNA does (*Villa-Komaroff* et al., 1975)? One would predict that DI-RNAs with translational capacity would be isolated. The other important property provided by the 5′ end sequences is probably the binding site for capsid proteins, which makes it possible to encapsidate the DI-RNA (*Bruton* et al., 1976). As is well-known, the 26 *S* RNA is not encapsidated during the normal infection and does not seem to bind capsid protein in the cells (*Ulmanen* et al., 1976). The possible role of defective interfering particles as a cause for chronic infections has been discussed by *Huang* and *Palma* (1975).

The reason for accumulation of DI particles is not known. It seems evident that DI-RNAs must be produced in every cycle of normal replication, possibly as mistakes in the synthesis of 26 *S* RNA. These DI-RNAs are encapsidated and obtain entrance to the cell. With the aid of the RNA polymerase of the standard virus, they replicate independently and their proportion increases during the further undiluted passages, causing interference in the replication of the standard virus RNA (*Huang,* 1973). As soon as the amount of standard virus becomes limiting, the replication of DI-RNAs stops. This type of interaction between DI particles and standard virus leads to fluctuation in the yields of standard virus and sometimes to an equilibrium yield of both particles, as has been described by *Johnston* et al. (1975) and *Guild* and *Stollar* (1975) (for VSV see *Huang,* 1973).

The role of host cells in the ability to replicate DI particles is evidently important, since DI particles isolated in vertebrate cells neither interfere nor stimulate synthesis of aberrant RNAs in *A. albopictus* cells (*Eaton,* 1975; *Igarashi* and *Stollar,* 1976). In murine cells, the dormant DI particles of the SF virus were readily replicated, causing interference and synthesis of DI-RNAs (*Levin* et al., 1973). These results suggest that recognition of the binding sites in the DI-RNAs may be regulated by host-specific components.

V. Conclusions

In this review, we have tried to outline the present knowledge about the structure and replication of the α-viruses. The different events of replication are schematically presented in Figure 11. It is obvious that the mechanism for many of the steps needs to be solved. The early events of interaction between host cells and α-viruses is so far poorly known. To date, neither the functions of the envelope glycoproteins nor the nature of the receptors has been determined. The isolated envelope proteins in their water soluble octamer form are presently used in the isolation of receptors, and hopefully this approach will solve the problem (*Helenius,* personal communication).

The mode of penetration and uncoating of α-viruses could be studied, e.g., by using radiolabeled virus and cell fractionation. Immunologic techniques combined with electron microscopy should show whether envelope proteins remain in the plasma membrane or are taken inside the cell.

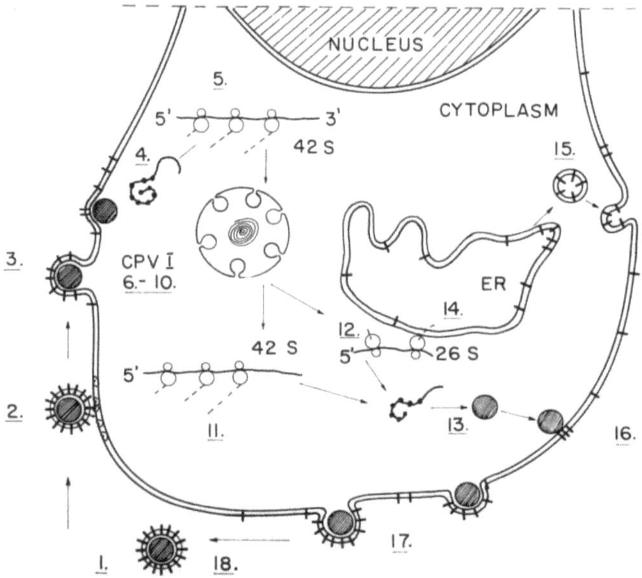

Fig. 11. Simplified scheme of α-virus replication. The virus (1) adsorbs to specific receptors (2) at the plasma membrane, fusing (?) with it (3). The nucleocapsid is released to the cytoplasm and uncoated (4). Ribosomes associate with the 42 *S* RNA genome, translating nonstructural proteins which are components of the RNA polymerase (*primary translation*, 5). When enough RNA polymerase has been assembled, the translation of the parental RNA is replaced by *primary transcription*, first producing 42 *S* RNA-negative strands (6) and by their transcription a first progeny of 42 *S* RNA-positive strands (7). These are used as messengers in the production of more nonstructural proteins during the *secondary translation* (8). Part of the positive strands are used parallely as templates for synthesis of more negative 42 *S* RNAs, which in turn are used as templates for synthesis of positive strands (*secondary transcription*, 9). When the concentration of the interconversion protein is high enough, the synthesis of 26 *S* RNA begins (10). The RNA synthesis probably takes place in cytoplasmic vacuoles (CPV I). The translation of 42 *S* RNA takes place in free polysomes (11), and the structural proteins are translated into membrane-bound polysomes (12). The progeny 42 *S* RNA and the capsid protein assemble into nucleocapsids (13). The envelope proteins are protruded into the cisternal side (14) of the endoplasmic reticulum membrane (ER), become glycosylated, first in ER and finally in the Golgi apparatus from which they are transported (15) to the plasma membrane. The nucleocapsid recognizes the spanning part of the envelope protein dimer (p62-E1), preventing its free lateral mobility (16). Increasing interactions between nucleocapsid and envelope proteins lead to protrusion of nucleocapsid into the plasma membrane (17). At this stage, the host cell proteins are excluded from the forming virion and p62 is cleaved to E2 and E3 (?). Finally, the mature virion is released into the medium (18)

The early events of protein synthesis remain a difficult problem because the viral synthesis is minimal in the beginning of infection compared to the host protein synthesis. The early phases of RNA synthesis can perhaps only be studied in vitro when the RNA polymerase components have been purified in active form. The mechanism of inhibition of host cell macromolecular synthesis has turned out to be a difficult task with all animal viruses so far.

Perhaps the use of temperature-sensitive mutants will help to solve these problems, as has been approached by *Atkins* (1976).

The shut off of the synthesis of nonstructural proteins late in the α-virus infection in favor of the structural proteins is an interesting problem. Do 42 *S* and 26 *S* RNA have different affinities to the ribosomes due to different nucleotide sequences at their 5′ ends or are there other factors involved? Actually, direct evidence is lacking that the initiation site for the structural proteins cannot be expressed in the 42 *S* RNA where this site is internal (*Cancedda* et al., 1975).

The translation and processing of the structural proteins still leave many unanswered questions; the nascent cleavage of capsid protein is probably essential for the further processing of the whole polyprotein. If the signal hypothesis of *Blobel* and *Dobberstein* (1975 a, b) is applicable for membrane glycoproteins, as it may be (*Katz* et al., 1977; *Wirth* et al., 1977), this cleavage is a necessary prerequisite for the release of the signal sequence which should immediately follow the capsid protein in the structural polyprotein. If the cleavage does not occur, the envelope proteins are not transported through the membrane.

Recently, the α-virus glycoproteins have been generally accepted as models for glycoprotein transport. There is fair hope that they will, together with the use of vesicular stomatitis virus G protein, help to establish the transport pathway of membrane glycoproteins from the site of translation to the plasma membrane. They will also provide models for the study of the glycosylation in different cells of vertebrate and invertebrate origin.

Finally, the analysis of the structure of the virus and especially the viral envelope by X-ray crystallographic methods has now become possible when proper crystals of the SF virus have been obtained (*Simons* et al., 1978). The detailed structural analysis of this virus with its homogenous piece of a biologic membrane is a tremendous task which will yield valuable information in the future.

Acknowledgements. We wish to express our gratitude to our colleagues *Carl-Henrik von Bonsdorff, Henrik Garoff, Niall Glanville, Ari Helenius, Sirkka Keränen, Kari Mattila, Marjut Ranki, Ossi Renkonen, Jaakko Saraste, Dorothea Sawicki, Kai Simons,* and *Ismo Ulmanen* for discussions and for giving us access to unpublished material. We are also indebted to Ms. *Päivi Lehtovuori* for expert secretarial aid and to Ms. *Satu Cankar* for drawing the graphs. The financial support from the Finnish Academy and the Sigrid Juselius foundation is gratefully acknowledged.

References

Aaslestad, H.G., Hoffman, E.J., Brown, A.: Fractionation of eastern equine encephalitis virus by density gradient centrifugation in CsCl. J. Virol. *2,* 972–978 (1968)

Abreu, S.L., Lucas-Lenard, J.: Cellular protein synthesis shut-off by Mengovirus: Translation of nonviral and viral mRNAs in extracts from uninfected and infected Ehrlich ascites cells. J. Virol. *18,* 183–194 (1976)

Acheson, N.H., Tamm, I.: Replication of Semliki Forest virus: An electron microscopic study. Virology *32,* 123–143 (1967)

Acheson, N.H., Tamm, I.: Purification and properties of Semliki Forest virus nucleocapsids. Virology *41,* 306–320 (1970a)

Acheson, N.H., Tamm, I.: Structural proteins of Semliki Forest virus and its nucleocapsid. Virology *41*, 321–329 (1970b)

Acheson, N.H., Tamm, I.: Ribonuclease sensivity of Semliki Forest virus nucleocapsids. J. Virol. *5*, 714–717 (1970c)

Agabalyan, A.S., Uryvaev, L.V., Ershov, F.I.: Characteristics of virion RNA of Venezuelan equine encephalomyelitis virus. Vopr. Virusol. *17*, 490–494 (1972)

Appleyard, G., Oram, J.D., Stanley, J.L.: Dissociation of Semliki Forest virus into biologically active components. J. Gen. Virol. *9*, 179–189 (1970)

Arif, B.M., Faulkner, P.: Genome of Sindbis virus. J. Virol. *9*, 102–109 (1971)

Armstrong, J.A., Edmonds, M., Nakazato, H., Phillips, B.A., Vaughan, M.H.: Polyadenylic acid sequences in the virion RNA of poliovirus and eastern equine encephalitis virus. Science *176*, 526–528 (1972)

Atkins, G.J.: The effect of infection with Sindbis virus and its temperature-sensitive mutants on cellular protein and DNA synthesis. Virology *71*, 593–597 (1976)

Atkins, G.J., Samuels, J., Kennedy, S.I.T.: Isolation and preliminary characterization of temperature-sensitive mutants of Sindbis virus strain AR 339. J. Gen. Virol. *25*, 371–380 (1974)

Baltimore, D., Girard, M.: An intermediate in the synthesis of poliovirus RNA. Proc. Natl. Acad. Sci USA *56*, 741–748 (1966)

Bancroft, J.B.: The self-assembly of spherical plant viruses. Adv. Virus Res. *16*, 99–133 (1970)

Becht, H., Rott, R., Klenk, H.D.: Effect of concanavalin A on cells infected with enveloped RNA viruses. J. Gen. Virol. *14*, 1–8 (1971)

Becker, R., Helenius, A., Simons, K.: Solubilization of the Semliki Forest virus membrane with sodium dodecyl sulfate. Biochemistry *14*, 1835–1841 (1975)

Bell, J.W., Waite, M.R.F.: Envelope antigens of Sindbis virus in cells infected with temperature-sensitive mutants. J. Virol. *21*, 788–791 (1977)

Ben-Ishai, Z., Goldblum, N., Becker, Y.: The intracellular site and sequence of Sindbis virus replication. J. Gen. Virol. *2*, 365–375 (1968)

Berget, S.M., Moore, C., Sharp, P.A.: Spliced segments at the 5′ terminus of adenovirus 2 late mRNA. Proc. Natl. Acad. Sci. USA *74*, 3171–3175 (1977)

Birdwell, C.R., Strauss, J.H.: Agglutination of Sindbis virus and of cells infected with Sindbis virus by plant lectins. J. Virol. *11*, 502–507 (1973)

Birdwell, C.R., Strauss, J.H.: Replication of Sindbis virus: IV. Electron microscope study of the insertion of viral glycoproteins into the surface of infected chick cells. J. Virol. *14*, 366–374 (1974a)

Birdwell, C.R., Strauss, J.H.: Distribution of the receptor sites for Sindbis virus on the surface of chicken and BHK cells. J. Virol. *14*, 672–678 (1974b)

Birdwell, C.R., Strauss, E.G., Strauss, J.H.: Replication of Sindbis virus: III. An electron microscopic study of virus maturation using the surface replica technique. Virology *56*, 429–438 (1973)

Blobel, G., Dobberstein, B.: Transfer of proteins across membranes. I. Presence of proteolytically processed and unprocessed nascent immunoglobulin light chains on membrane-bound ribosomes of murine myeloma. J. Cell Biol. *67*, 835–851 (1975a)

Blobel, G., Dobberstein, B.: Transfer of proteins across membranes. II. Reconstitution of functional rough microsomes from heterologous components. J. Cell Biol. *67*, 852–862 (1975b)

Bonsdorff, C.-H. v.: Structural role of RNA in Semliki Forest virus nucleocapsid. Acta Pathol. Microbiol. Scand. Sect B *80*, 579–588 (1972)

Bonsdorff, C.-H. v.: The structure of Semliki Forest virus. Commentationes Biologie *74*, 1–53 (Societas Scientarium Fennica, Helsinki, Finland) (1973)

Bonsdorff, C.-H., Harrison, S.C.: Sindbis virus glycoproteins form a regular icosahedral surface lattice. J. Virol. *16*, 141–145 (1975)

Bose, H.R., Sagik, B.P.: The virus envelope in cell attachment. J. Gen. Virol. *9*, 159–161 (1970)

Boulton, R.W., Westaway, E.G.: Comparisons of togaviruses: Sindbis virus (group A) and Kunjin virus (group B). Virology *49*, 283–289 (1972)

Bracha, M., Schlesinger, M.J.: Inhibition of Sindbis virus replication by zinc ions. Virology 72, 272–277 (1976a)

Bracha, M., Schlesinger, M.J.: Defects in RNA$^+$ temperature-sensitive mutants of Sindbis virus and evidence for a complex of PE2-E1 viral glycoproteins. Virology 74, 441–449 (1976b)

Bracha, M., Leone, A., Schlesinger, M.J.: Formation of a Sindbis virus nonstructural protein and its relation to 42S mRNA function. J. Virol. 20, 612–620 (1976)

Bracha, M., Sagher, D., Brown, A., Schlesinger, M.J.: The protease inhibitor p-Nitrophenyl-p-Guanidine-benzoate inactivates Sindbis and other enveloped viruses. Virology 77, 45–55 (1977)

Bras-Herreng, F.: Multiplication of Sindbis virus in Drosophila cells cultivated in vitro. Arch. Virol. 48, 121–130 (1975)

Brawner, T.A., Lee, J.C., Trent, D.W.: A comparison of Saint Louis encephalitis and Sindbis virus RNA. Arch. Virol. 54, 147–151 (1977)

Brotherus, J., Renkonen, O.: Subcellular distribution of lipids in BHK cells: evidence for the enrichment of lysobis phosphatidic acid and neutral lipids in lysosomes. J. Lipid. Res. 18, 191–202 (1977)

Brown, D.T., Gliedman, J.B.: Morphological variants of Sindbis virus obtained from infected mosquito tissue culture cells. J. Virol. 12, 1534–1539 (1973)

Brown, D.T., Smith, J.F.: Morphology of BHK-21 cells infected with Sindbis virus temperature-sensitive mutants in complementation groups D and E. J. Virol. 15, 1262–1266 (1975)

Brown, D.T., Waite, M.R.F., Pfefferkorn, E.R.: Morphology and morphogenesis of Sindbis virus as seen with freeze-etching techniques. J. Virol. 10, 524–536 (1972)

Brown, F., Smale, C.J., Horzinek, M.C.: Lipid protein organization in vesicular stomatitis and Sindbis viruses. J. Gen. Virol. 23, 455–458 (1973)

Bruton, C.J., Kennedy, S.I.T.: Semliki Forest virus intracellular RNA: Properties of the multistranded RNA species and kinetics of positive and negative strand synthesis. J. Gen. Virol. 28, 111–128 (1975)

Bruton, C.J., Kennedy, S.I.T.: Defective-interfering particles of Semliki Forest virus: Structural differences between standard virus and defective-interfering particles. J. Gen. Virol. 31, 383–395 (1976)

Bruton, C.J., Porter, A., Kennedy, S.I.T.: Defective-interfering particles of Semliki Forest virus: Intracellular events during interference. J. Gen. Virol. 31, 397–416 (1976)

Brzeski, H., Kennedy, S.I.T.: Synthesis of Sindbis virus nonstructural polypeptides in chicken embryo fibroblasts. J. Virol. 22, 420–429 (1977)

Buckley, S.M.: Susceptibility of the *Aedes albopictus* and *Aedes aegypti* cell lines to infection with arboviruses. Proc. Soc. Exp. Biol. Med. 131, 625–630 (1969)

Burge, B.W., Huang, A.S.: Comparison of membrane protein glycopeptides of Sindbis virus and vesicular stomatitis virus. J. Virol. 6, 176–182 (1970)

Burge, B.W., Pfefferkorn, E.R.: Isolation and characterization of conditional-lethal mutants of Sindbis virus. Virology 30, 204–213 (1966a)

Burge, B.W., Pfefferkorn, E.R.: Complementation between temperature-sensitive mutants of Sindbis virus. Virology 30, 214–223 (1966b)

Burge, B.W., Pfefferkorn, E.R.: Phenotypic mixing between group A arboviruses. Nature (London) 210, 1397–1399 (1966c)

Burge, B.W., Pfefferkorn, E.R.: Temperature-sensitive mutants of Sindbis virus: Biochemical correlations of complementation. J. Virol. 1, 956–962 (1967)

Burge, B.W., Pfefferkorn, E.R.: Functional defects of temperature-sensitive mutants of Sindbis virus. J. Mol. Biol. 35, 193–205 (1968)

Burge, B.W., Strauss, J.H.: Glycopeptides of the membrane glycoprotein of Sindbis virus. J. Mol. Biol. 47, 449–466 (1970)

Burke, D.C.: Processing of alphavirus proteins in infected cells. Med. Biol. 53, 352–356 (1975)

Burke, D.J., Keegstra, K.: Purification and composition of the proteins from Sindbis virus grown in chick and BHK cells. J. Virol. 20, 676–686 (1976)

Butler, P.J.G.: The mechanism and control of the assembly of Tobacco Mosaic virus from its RNA and protein disks. Cold Spring Harbor Symp. Quant. Biol. *36*, 461–468 (1971)

Bykovsky, A.F., Yershov, F.I., Zhdanov, V.M.: Morphogenesis of Venezuelan equine encephalomyelitis virus. J. Virol. *4*, 496–504 (1969)

Cancedda, R., Schlesinger, M.J.: Formation of Sindbis virus capsid protein in mammalian cell-free extracts programmed with viral messenger RNA. Proc. Natl. Acad. Sci. USA *71*, 1843–1847 (1974)

Cancedda, R., Swanson, R., Schlesinger, M.J.: Effects of different RNAs and components of the cell-free system on in vitro synthesis of Sindbis viral proteins. J. Virol. *14*, 652–663 (1974a)

Cancedda, R., Swanson, R., Schlesinger, M.J.: Viral proteins formed in a cell-free rabbit reticulocyte system programmed with RNA from a temperature-sensitive mutant of Sindbis virus. J. Virol. *14*, 664–671 (1974b)

Cancedda, R., Villa-Komaroff, L., Lodish, H.F., Schlesinger, M.J.: Initiation sites for transcription of Sindbis virus 42S and 26S messenger RNAs. Cell *6*, 215–222 (1975)

Calberg-Bacq, C.M., Osterrieth, P.M.: Morphological modifications of Semliki Forest virus after treatment with pronase. Acta Virol. Prague *10*, 266–267 (1966)

Cartwright, K.L., Burke, D.C.: Virus nucleic acids formed in chick embryo cells infected with Semliki Forest virus. J. Gen. Virol. *6*, 231–248 (1970)

Casals, J.: International arbovirus research. Med. Biol. *53*, 249–258 (1975)

Casals, J., Clarke, D.H.: Arboviruses: Group A. In: Viral and Ricketsial Infections of Man. *Horsfall, F.L., Tamm, I.* (Eds.) London: Pitman Medical Publishing Co., LTD, 1965, pp. 583–605

Chain, M.M.T., Doane, F.W., McLean, D.M.: Morphological development of Chicungunya virus. Can. J. Microbiol. *12*, 895–900 (1966)

Cheng, P.-Y.: Infectivity of ribonucleic acid from mouse brains infected with Semliki Forest virus. Nature (London) *181*, 1800 (1958)

Cheng, P.-Y.: Purification, size, and morphology of a mosquito-borne animal virus, Semliki Forest virus. Virology *14*, 124–131 (1961)

Choppin, P.W., Compans, R.W.: Phenotypic mixing of envelope proteins of the parainfluenza virus SV5 and vesicular stomatitis virus. J. Virol. *5*, 609–616 (1970)

Chow, L.T., Gelinas, R.E., Broker, T.R., Roberts, R.J.: An amazing sequence arrangement at the 5′ ends of adenovirus 2 messenger RNA. Cell *12*, 1–8 (1977)

Clegg, J.C.S.: Sequential translation of capsid and membrane protein genes in arbovirus-infected cells. Nature (London) *254*, 454–455 (1975)

Clegg, J.C.S., Kennedy, S.I.T.: Polyadenylic acid sequences in the virus RNA species of cells infected with Semliki Forest virus. J. Gen. Virol. *22*, 331–345 (1974a)

Clegg, J.C.S., Kennedy, S.I.T.: In vitro synthesis of structural proteins of Semliki Forest virus directed by isolated 26S RNA from infected cells. FEBS Lett. *42*, 327–330 (1974b)

Clegg, C., Kennedy, I.: Translation of Semliki-Forest-virus intracellular 26-S RNA: Characterisation of the products synthesized in vitro. Eur. J. Biochem. *53*, 175–184 (1975a)

Clegg, J.C.S., Kennedy, S.I.T.: Initiation of synthesis of the structural proteins of Semliki Forest virus. J. Mol. Biol. *97*, 401–411 (1975b)

Clegg, J.C.S., Kennedy, S.I.T.: Translation of the genes for the structural proteins of alphaviruses. Med. Biol. *53*, 383–386 (1975c)

Clegg, J.C.S., Brzeski, H., Kennedy, S.I.T.: RNA polymerase components in Semliki Forest virus infected cells: Synthesis from large precursors. J. Gen. Virol. *32*, 413–430 (1976)

Clewley, J.P., Kennedy, S.I.T.: Purification and polypeptide composition of Semliki Forest virus RNA polymerase. J. Gen. Virol. *32*, 395–411 (1976)

Colbere, F., Hannoun, C.: Viral specific RNA synthesis in Sindbis infected Aedes albopictus mosquito cells: I. Evidence for a 18S RNA species in infected cells. Ann. Microbiol. *125B*, 393–406 (1974)

Cole, C.N., Baltimore, D.: Defective interfering particles of poliovirus. IV. Mechanism of enrichment. J. Virol. *12*, 1414–1426 (1973)

Compans, R.W.: Location of the glycoprotein in the membrane of Sindbis virus. Nature New Biol. *229*, 114–116 (1971)

Dalrymple, J.M., Schlesinger, S., Russell, P.K.: Antigenic characterization of two Sindbis envelope glycoproteins separated by isoelectric focusing. Virology *69*, 93–103 (1976)

Dalrymple, J.M., Vogel, S.N., Teramoto, A.Y., Russell, P.K.: Antigenic components of group A arbovirus virions. J. Virol. *12*, 1034–1042 (1973)

Davey, M.W., Dalgarno, L.: Semliki Forest virus replication in cultured Aedes albopictus cells: Studies on the establishment of persistence. J. Gen. Virol. *24*, 453–463 (1974)

Davey, M.W., Dennett, D.P., Dalgarno, L.: The growth of two togaviruses in cultured mosquito and vertebrate cells. J. Gen. Virol. *20*, 225–232 (1973)

David, A.E.: Lipid composition of Sindbis virus. Virology *46*, 711–720 (1971)

Deborde, D.C., Leibowitz, R.D.: Polyadenylic acid size and position found in Sindbis virus genome and mRNA species. Virology *72*, 80–88 (1976)

Dobberstein, B.: Signals for the Vectorial Transfer of Proteins in the Cell. Reported at EMBO Workshop On Membrane Proteins, Helsinki, 1977

Dobos, P., Faulkner, P.: Characterization of a cytoplasmic fraction from Sindbis virus-infected cultures. Can. J. Microbiol. *15*, 215–222 (1968)

Dobos, P., Faulkner, P.: Properties of 42S and 26S Sindbis viral RNA species. J. Virol. *4*, 429–438 (1969)

Dobos, P., Faulkner, P.: On the infectivity of the Sindbis virus nucleocapsid. Can. J. Microbiol. *16*, 1273–1283 (1970a)

Dobos, P., Faulkner, P.: Molecular weight of Sindbis virus ribonucleic acid as measured by polyacrylamide gel electrophoresis. J. Virol. *6*, 145–147 (1970b)

Dobos, P., Arif, B.M., Faulkner, P.: Denaturation of Sindbis virus RNA with dimethyl sulphoxide. J. Gen. Virol. *10*, 102–106 (1971)

Dubin, D.T., Stollar, V.: Methylation of Sindbis "26S" messenger RNA. Biochem. Biophys. Res. Commun. *66*, 1373–1379 (1975)

Dubin, D.T., Stollar, V., Hsuchen, C.-C., Timko, K., Guild, G.M.: Sindbis virus messenger RNA: The 5'-termini and methylated residues of 26 and 42S RNA. Virology *77*, 457–470 (1977)

Duda, E., Schlesinger, M.J.: Alterations in Sindbis viral envelope proteins by treating BHK cells with glucosamine. J. Virol. *15*, 416–419 (1975)

Dulbecco, R., Vogt, M.: One-step growth curve of western equine encephalitis virus on chicken embryo cells grown in vitro and analysis of virus yielded from single cells. J. Exp. Med. *99*, 183–199 (1954)

Eaton, B.T.: Defective interfering particles of SFV generated in BHK cells do not interfere with viral RNA synthesis in aedes albopictus cells. Virology *68*, 534–538 (1975)

Eaton, B.T.: Evidence for the synthesis of defective interfering particles by aedes albopictus cells persistently infected with Sindbis virus. Virology *77*, 843–848 (1977)

Eaton, B.T., Faulkner, P.: Heterogeneity in the poly (A) content of the genome of Sindbis virus. Virology *50*, 865–873 (1972)

Eaton, B.T., Faulkner, P.: Altered pattern of viral RNA synthesis in cells infected with standard and defective Sindbis virus. Virology *51*, 85–93 (1973)

Eaton, B.T., Hapel, A.J.: Persistent noncytolytic togavirus infection of primary mouse muscle cells. Virology *72*, 266–271 (1976)

Eaton, B.T., Regnery, R.L.: Polysomal RNA in Semliki Forest virus infected Aedes albopictus cells. J. Gen. Virol. *29*, 35–50 (1975)

Eaton, B.T., Donaghue, T.P., Faulkner, P.: Presence of poly (A) in the polyribosome associated RNA of Sindbis-infected BHK cells. Nature New Biol. *238*, 109–111 (1972)

Engelhardt, D.L.: Assay for secondary structure in ribonucleic acid. J. Virol. *9*, 903–908 (1972)

Erlandson, R.A., Babcock, V.I., Southam, C.M., Bailey, R.B., Shipkey, F.H.: Semliki Forest virus in HEp 2 cell cultures. J. Virol. *1*, 996–1009 (1967)

Ershov, F.I., Men'Shikh, L.K., Zaitseva, O.V., Zhdanov, V.M.: RNA-polymerase induced by Venezuelan equine encephalomyelitis virus. Voz. Virusol. *16*, 103–108 (1971)

Esparza, J., Sanches, A.: Multiplication of Venezuelan equine encephalitis (Mucambo) virus in cultured mosquito cells. Arch. Virol. *49*, 273–280 (1975)

Faulkner, P., McGee-Russel, S.M.: Purification and structure of Semliki Forest virus isolated from mouse brain (baby hamster kidney cells). Can. J. Microbiol. *14*, 153–159 (1968)

Fenner, F.: Classification and nomenclature of viruses: Second report of the international committee on taxonomy of viruses. Intervirology 7, 1–115 (1976)

Follet, E.A.C., Pringle, C.R., Pennington, T.H.: Virus development in enucleate cells: Echo-

virus, poliovirus, pseudorabies virus, reovirus, respiratory syncytial virus and Semliki
Forest virus. J. Gen. Virol. *26*, 183–196 (1975)

Friedman, R.M.: Protein synthesis directed by an arbovirus. J. Virol. *2*, 26–32 (1968a)

Friedman, R.M.: Replicative intermediate of an arbovirus. J. Virol. *2*, 547–552 (1968b)

Friedman, R.M.: Structural and nonstructural proteins of an arbovirus. J. Virol. *2*,
1076–1080 (1968c)

Friedman, R.M., Berezesky, I.K.: Cytoplasmic fractions associated with Semliki Forest
virus ribonucleic acid replication. J. Virol. *1*, 274–383 (1967)

Friedman, R.M., Grimley, P.M.: Inhibition of Arbovirus assembly by cycloheximide. J.
Virol. *4*, 292–299 (1969)

Friedman, R.M., Pastan, I.: Nature and function of the structural phospholipids of an
arbovirus. J. Mol. Biol. *40*, 107–115 (1969)

Friedman, R.M., Sonnabend, J.A.: Inhibition by interferon of production of double-stranded
Semliki Forest virus ribonucleic acid. Nature (London) *206*, 532 (1965)

Friedman, R.M., Sreevalsan, T.: Membrane binding of input arbovirus ribonucleic acid:
Effect of interferon or cycloheximide. J. Virol. *6*, 169–175 (1970)

Friedman, R.M., Fantes, K.H., Levy, H.B., Carter, W.B.: Interferon action on parental
Semliki Forest virus ribonucleic acid. J. Virol. *1*, 1168–1173 (1967)

Friedman, R.M., Levin, J.G., Grimley, P.M., Berezesky, I.K.: Membrane-associated replication complex in arbovirus infection. J. Virol. *10*, 504–515 (1972)

Friedman, R.M., Levy, H.B., Carter, W.B.: Replication of Semliki Forest virus: Three
forms of viral RNA produced during infection. Proc. Natl. Acad. Sci. USA *56*, 440–446
(1966)

Frish-Niggemeyer, W.: Die chemische Struktur der bei der Adsorption von TBE-virus
an Erythrozyten wirksamen Rezeptorsubstanz. Arch. Hyg. Bakteriol. *151*, 585–598 (1967)

Froger, C., Louisot, P.: Glycoprotein biosynthesis in arbovirus-infected cells: I. Study of
glucosamine and N-acetyl-glucosamine transferases. Comp. Biochem. Physiol. B Comp.
Biochem. *43*, 223–231 (1972a)

Froger, C., Louisot, P.: Glycoprotein biosynthesis in arbovirus infected cells: II. Study
of microsomic mannosyl transferase activity. Int. J. Biochem. *3*, 613–622 (1972b)

Fuscaldo, A.A., Aaslestad, H.G., Hoffman, E.J.: Biological physical and chemical properties
of Eastern equine encephalitis virus. J. Virol. *7*, 233–241 (1971)

Gahmberg, C.G., Simons, K., Renkonen, O., Kääriäinen, L.: Exposure of proteins and
lipids in the Semliki Forest virus membrane. Virology *50*, 259–262 (1972a)

Gahmberg, C.G., Uterman, G., Simons, K.: The membrane proteins of Semliki Forest
virus have a hydrophobic part attached to the viral membrane. FEBS Lett. *28*, 179–182
(1972b)

Garoff, H.: Cross-linking of the spike glycoproteins in Semliki Forest virus with dimethyl-
suberimidate. Virology *62*, 385–392 (1974)

Garoff, H., Simons, K.: Location of the spike glycoproteins in the Semliki Forest virus
membrane. Proc. Natl. Acad. Sci. USA *71*, 3988–3992 (1974)

Garoff, H., Söderlund, H.: The amphiphilic membrane glycoproteins of Semliki Forest
virus are attached to the lipid bilayer by their COOH-terminal end. J. Mol. Biol.
in press (1978)

Garoff, H., Simons, K., Renkonen, O.: Isolation and characterisation of the membrane
proteins of Semliki Forest virus. Virology *61*, 493–504 (1974)

Gelinas, R.C., Roberts, R.J.: One predominant 5′ undeca nucleotide in adenovirus 2 late
messenger RNAs. Cell *11*, 533–544 (1977)

Gil-Fernandez, C., Ronda-Lain, C., Rubiohuertos, M.: Electron microscopic study of Sindbis
virus morphogenesis. Arch. Ges. Virusforsch. *40*, 1–9 (1973)

Glanville, N., Lachmi, B.: Translation of proteins accounting for the full coding capacity
of the Semliki Forest virus 42S RNA genome. FEBS Lett. *81*, 399–402 (1977)

Glanville, N., Ulmanen, I.: Biological activity of in vitro synthesized protein: Binding
of Semliki Forest virus capsid protein to the large ribosomal subunit. Biochem. Biophys.
Res. Commun. *71*, 393–399 (1976)

Glanville, N., Lachmi, B., Smith, A.E., Kääriäinen, L.: Tryptic peptide mapping of the nonstructural proteins of Semliki Forest virus. Biochim. Biophys. Acta *518*, 497–506 (1978)

Glanville, N., Morser, J., Uomala, P., Kääriäinen, L.: Simultaneous translation of structural and nonstructural proteins from Semliki Forest virus RNA in two eukaryotic systems in vitro. Eur. J. Biochem. *64*, 167–175 (1976a)

Glanville, N., Ranki, M., Morser, J., Kääriäinen, L., Smith, A.E.: Initiation of translation directed by 42S and 26S RNAs from Semliki Forest virus in vitro. Proc. Natl. Acad. Sci. USA *73*, 3059–3063 (1976b)

Gliedman, J.B., Smith, J.F., Brown, D.T.: Morphogenesis of Sindbis virus in cultured Aedes albopictus cells. J. Virol. *16*, 913–926 (1975)

Gorman, B.: Lipid inhibitors of arbovirus haemagglutination. J. Gen. Virol. *6*, 305–313 (1970)

Grimes, W.J., Burge, B.W.: Modification of Sindbis virus glycoprotein by host-specified glycosyl transferases. J. Virol. *7*, 309–313 (1971)

Grimley, P.M., Friedman, R.M.: Development of Semliki Forest virus in mouse brain: An electron microscopic study. Exp. Mol. Pathol. *12*, 1–13 (1970)

Grimley, P.M., Berezesky, I.K., Friedman, R.M.: Cytoplasmic structures associated with an arbovirus infection: loci of viral ribonucleic acid synthesis. J. Virol. *2*, 1326–1338 (1968)

Grimley, P.M., Levin, J.G., Berezesky, I.K., Friedman, R.M.: Specific membranous structures associated with the replication of group A arboviruses. J. Virol. *10*, 492–503 (1972)

Guild, G.M., Stollar, V.: Defective interfering particles of Sindbis virus: III. Intracellular viral RNA species in chick embryo cell cultures. Virology *67*, 24–41 (1975)

Guild, G.M., Stollar, V.: Defective interfering particles of Sindbis virus: V. Sequence relationships between SV$_{std}$ 42S RNA and intracellular defective viral RNAs. Virology *77*, 175–188 (1977)

Guild, G.M., Flores, L., Stollar, V.: Defective interfering particles of Sindbis virus: IV. Virion RNA species and molecular weight determination of defective double-stranded RNA. Virology *77*, 158–174 (1977)

Haahtela, K., Renkonen, O.: The structure of "core tetra saccharide" of Semliki Forest E3 glycoprotein. Manuscript in preparation. (1978)

Hammer, G., Schwarz, R.T., Scholtissek, C.: Effect of infection with enveloped viruses on nucleotide metabolism. Virology *70*, 238–240 (1976)

Hardy, F.M., Brown, A.: Growth of Venezuelan equine encephalomyelitis virus in L cells. II. Growth in submerged culture. J. Bacteriol. *82*, 449–457 (1961)

Harrison, S.C., David, A., Jumblatt, J., Darnell, J.E.: Lipid and protein organization in Sindbis virus. J. Mol. Biol. *60*, 523–528 (1971)

Harrison, S.C., Jack, A., Goodenough, D., Sefton, B.M.: Structural studies of sperical viruses. J. Supramol. Struct. *2*, 486–495 (1974)

Hashimoto, K., Suzuki, K., Simizu, B.: Morphological and physical properties of a multiploid-forming mutant of western equine encephalitis virus. J. Virol. *15*, 1454–1466 (1975)

Hashimoto, K., Suzuki, K., Simizu, B.: Maturation defect of a temperature-sensitive mutant of Western equine encephalitis virus. Arch. Virol. *53*, 209–219 (1977)

Hay, A.J., Skehel, J.J., Burke, D.C.: Proteins synthesised in chick cells following infection with Semliki Forest virus. J. Gen. Virol. *3*, 175–184 (1968)

Hefti, E., Bishop, D.H.L., Dubin, D.T., Stollar, V.: 5′ nucleotide sequence of Sindbis viral RNA. J. Virol. *17*, 149–159 (1976)

Helenius, A., Bonsdorff, C.-H. v.: Semliki Forest virus membrane proteins: Preparation and characterization of spike complexes soluble in detergent-free medium. Biochim. Biophys. Acta *436*, 895–899 (1976)

Helenius, A., Simons, K.: The binding of detergents to lipophilic and hydrophilic proteins. J. Biol. Chem. *247*, 3656–3661 (1972)

Helenius, A., Simons, K.: Solubilization of membranes by detergents. Biochim. Biophys. Acta *415*, 29–79 (1975)

Helenius, A., Söderlund, H.: Stepwise dissociation of the Semliki Forest virus membrane with Triton X-100. Biochim. Biophys. Acta *307*, 287–300 (1973)

Helenius, A., Fries, E., Garoff, H., Simons, K.: Solubilization of the Semliki Forest virus membrane with sodium deoxycholate. Biochim. Biophys. Acta *436*, 319–334 (1976)

Heydrick, F.P., Comer, J.F., Wachter, R.F.: Phospholipid composition of Venezuelan equine encephalitis virus. J. Virol. *7*, 642–645 (1971)

Higashi, N.: Electron microscopy of the multiplication of Chikungunya virus in cell cultures. Japn. J. Southeast Asian Stud. *4*, 88–94 (1966)

Higashi, N.: Electron microscopy of viruses in thin sections of cells grown in culture. Prog. Med. Virol. *15*, 331–379 (1973)

Higashi, N., Matsumoto, A., Tabata, K., Nagamoto, Y.: Electron microscope study of development of Chikungunya virus in green monkey kidney stable (VERO) cells. Virology *33*, 55–69 (1967)

Hirano, H., Parkhouse, B., Nicolson, G.L., Lennox, E.S., Singer, S.I.: Distribution of saccharide residues on membrane fragments from a myeloma-cell homogenate: its implications for membrane biogenesis. Proc. Natl. Acad. Sci. USA *69*, 2945–2949 (1972)

Hirschberg, C.B., Robbins, P.W.: The glycolipids and phospholipids of Sindbis virus and their relation to the lipids of the host cell plasma membrane. Virology *61*, 602–608 (1974)

Horzinek, M.C.: The structure of togaviruses. Prog. Med. Virol. *16*, 109–156 (1973a)

Horzinek, M.C.: Comparative aspects of Togaviruses. J. Gen. Virol. *20*, 87–103 (1973b)

Horzinek, M.C.: The structure of Togaviruses and Bunyaviruses. Med. Biol. *53*, 406–411 (1975)

Horzinek, M., Mussgay, M.: Studies on the nucleocapsid structure of a group A arbovirus. J. Virol. *4*, 514–520 (1969)

Horzinek, M., Mussgay, M.: Studies on the substructure of togaviruses: I. Effect of urea, deoxycholate, and saponin on the Sindbis virion. Arch. Ges. Virusforsch. *33*, 296–305 (1971)

Hsu, M.-T., Kung, H.-J., Davidson, N.: An electron microscopic study of Sindbis virus RNA. Cold Spring Harbor Symp. Quant. Biol. *38*, 943–950 (1973)

Hsuchen, C.-C., Dubin, D.T.: Di- and trimethylated congeners of 7-methyl guanine in Sindbis virus mRNA. Nature (London) *264*, 190–191 (1976)

Huang, A.S.: Defective interfering viruses. Ann. Rev. Microbiol. *27*, 101–117 (1973)

Huang, A.S., Palma, E.L.: Defective interfering particles as antiviral agents. Perspectives in Virology *9*, 77–90 (1975)

Hughes, F., Pedersen, C.E.: Paramagnetic spin label interactions with the envelope of a group A arbovirus: lipid organization. Biochim. Biophys. Acta *394*, 102–110 (1975)

Igarashi, A.: Protein synthesis and formation of Chikungunya virus in infected BHK21 cells. Biken J. *13*, 289–302 (1970)

Igarashi, A., Stollar, V.: Failure of defective interfering particles of Sindbis virus produced in BHK or chicken cells to affect viral replication in Aedes albopictus cells. J. Virol. *19*, 398–408 (1976)

Igarashi, A., Fukuoka, T., Nithiuthai, P., Hsu, L.-C., Fukai, K.: Structural components of Chikungunya virus. Biken J. *13*, 93–110 (1970)

Igarashi, A., Koo, R., Stollar, V.: Evolution and properties of Aedes albopictus cell cultures persistently infected with Sindbis virus. Virology *82*, 69–83 (1977)

Igarashi, A., Sasao, F., Fukai, K.: Intracellular components associated with Chikungunya virus-specific ribonucleic acids in infected BHK21 cells. Japn. J. Microbiol. *17*, 401–408 (1973)

Inglot, A.D., Albin, M., Chudzio, T.: Persistent infection of mouse cells with Sindbis virus: Role of virulence of strains, auto-interfering particles and interferon. J. Gen. Virol. *20*, 105–110 (1973)

Ivanic, S.: Identification of two envelope proteins of Semliki Forest virus before and after treatment with Triton X-100. Arch. Ges. Virusforsch. *44*, 164–167 (1974)

Johnson, I., Clamp, J.R.: The oligosaccharides of human type L immunoglobulin M (macroglobulin). Biochem. J. *123*, 739–745 (1971)

Johnson, R.T.: Virus invasion of the central nervous system: A study of Sindbis virus infection in the mouse using fluorescent antibody. Am. J. Pathol. *46*, 929–943 (1965)

Johnston, R.E., Bose, H.R.: An adenylate-rich segment in the virion RNA of Sindbis virus. Biochem. Biophys. Res. Commun. *46*, 712–718 (1972)

Johnston, R.E., Tovell, D.R., Brown, D.T., Faulkner, P.: Interfering passages of Sindbis virus: Concomitant appearance of interference, morphological variants, and truncated viral RNA. J. Virol. *16*, 951–958 (1975)

Jones, K.H., Waite, M.R.F., Bose, H.R.: Cleavage of a viral envelope precursor during the morphogenesis of Sindbis virus. J. Virol. *13*, 809–817 (1974)

Jones, K.J., Scupham, R.K., Pfeil, J.A., Wan, K., Sagik, B.P., Bose, H.R.: Interaction of Sindbis virus glycoproteins during morphogenesis. J. Virol. *21*, 778–787 (1977)

Kääriäinen, L., Gomatos, P.J.: A kinetic analysis of the synthesis in BHK 21 cells of RNAs specific for Semliki Forest virus. J. Gen. Virol. *5*, 251–265 (1969)

Kääriäinen, L., Renkonen, O.: Envelopes of lipid-containing viruses as models for membrane assembly. In: Cell Surface Reviews. Vol. 4. Poste, G., Nicolson, G.L. (Eds.) Elsevier/ North-Holland Biomedical Press, 1977, pp. 741–801

Kääriäinen, L., Söderlund, H.: Properties of Semliki Forest virus nucleocapsid: Sensitivity to pancreatic ribonuclease. Virology *43*, 291–299 (1971)

Kääriäinen, L., Glanville, N., Keränen, S., Lachmi, B., Morser, J., Ranki, M., Uomala, P.: Translation of Semliki Forest virus RNAs in vivo and in vitro. Colloques de l'INSERM *47*, 265–272 (1975a)

Kääriäinen, L., Keränen, S., Lachmi, B., Söderlund, H., Tuomi, K., Ulmanen, I.: Replication of Semliki Forest virus. Med. Biol. *53*, 342–352 (1975b)

Kääriäinen, L., Lachmi, B., Glanville, N.: Translational control in Semliki Forest virus infected cells. Ann. Microbiol. (Inst. Pasteur) *127A*, 197–203 (1976)

Kääriäinen, L., Sawicki, D., Gomatos, P.J.: Cleavage defect in the nonstructural polyprotein of Semliki Forest virus has two separate effects on viral RNA synthesis. J. Gen. Virol. *39*, 463–473 (1978)

Kääriäinen, L., Simons, K., Bonsdorff, C.-H. v.: Studies in subviral components of Semliki Forest virus. Ann. Med. Exp Biol. Fenn. *47*, 235–248 (1969)

Kaluza, G.: Effect of impaired glycosylation on the biosynthesis of Semliki Forest virus glycoproteins. J. Virol. *16*, 602–612 (1975)

Kaluza, G.: Early synthesis of Semliki Forest virus-specific proteins in infected chicken cells. J. Virol. *19*, 1–12 (1976)

Kaluza, G., Kraus, A.A., Rott, R.: Inhibition of cellular protein synthesis by simultaneous pretreatment of host cells with fowl plaque virus and actinomycin D.: a method for studying early protein synthesis of several RNA viruses. J. Virol. *17*, 1–9 (1976)

Kaluza, G., Schmidt, M.F.G., Scholtissek, C.: Effect of 2-deoxy-D-glucose on the multiplication of Semliki Forest virus and the reversal of the block by mannose. Virology *54*, 179–189 (1973)

Kaper, J.M.: Arrangement and identification of simple isometric viruses according to their dominating stabilizing interactions. Virology *55*, 299–304 (1973)

Kaper, J.M.: In: The Chemical Basis of Virus Structure. Dissociation and reassembly. Amsterdam: North-Holland, 1975

Karabatsos, N.: Density gradient characterization of disrupted and untreated eastern encephalitis virus. Arch. Ges. Virusforsch. *40*, 222–235 (1973)

Katz, F.N., Rothman, J.E., Lingappa, V.R., Blobel, G., Lodish, H.F.: Membrane assembly in vitro: Synthesis, glycosylation and assymmetric insertion of a transmembrane protein. Proc. Natl. Acad. Sci. USA *74*, 3278–3282 (1977)

Keegstra, K., Sefton, B., Burke, D.: Sindbis virus glycoproteins: Effect of the host cell on the oligosaccharides. J. Virol. *16*, 613–620 (1975)

Kennedy, S.I.T.: Isolation and identification of the virus-specified RNA species found on membrane-bound polyribosomes of chick embryo cells infected with Semliki Forest virus. Biochem. Biophys. Res. Commun. *48*, 1254–1258 (1972)

Kennedy, S.I.T.: The effect of enzymes on structural and biological properties of Semliki Forest virus. J. Gen. Virol. *23*, 129–143 (1974)

Kennedy, S.I.T.: Sequence relationships between the genome and the intracellular RNA species of standard and defective-interfering Semliki Forest virus. J. Mol. Biol. *108*, 491–511 (1976)

Kennedy, S.I.T., Burke, D.C.: Studies on the structural proteins of Semliki Forest virus. J. Gen. Virol. *14*, 87–98 (1972)

Kennedy, S.I.T., Bruton, C.J., Weiss, B., Schlesinger, S.: Defective interfering passages of Sindbis virus: Nature of the defective virion RNA. J. Virol. *19*, 1034–1043 (1976)

Keränen, S.: Complementation in 26S RNA synthesis between temperature-sensitive mutants of Semliki Forest virus. FEBS Lett. *80*, 164–168 (1977a)

Keränen, S.: Interference of wild type virus replication by an RNA negative temperature-sensitive mutant of Semliki Forest virus. Virology *80*, 1–11 (1977b)

Keränen, S.: Studies on the replication of Semliki Forest virus using temperature sensitive mutants. Ph.D. Thesis, University of Helsinki, Helsinki, Finland (1977c)

Keränen, S., Kääriäinen, L.: Isolation and basic characterization of temperature-sensitive mutants from Semliki Forest virus. Acta Pathol. Microbiol. Scand. Sect. B *82*, 810–820 (1974)

Keränen, S., Kääriäinen, L.: Proteins synthesized by Semliki Forest virus and its 16 temperature-sensitive mutants. J. Virol. *16*, 388–396 (1975)

Klessig, D.F.: Two adenovirus mRNAs have a common 5′ terminal leader sequence encoded at least 10 kb upstream from their main coding regions. Cell *12*, 9–21 (1977)

Koblett, H., Wittek, R., Kohler, U., Wyler, R.: Ribonucleic acid of Semliki Forest virus. Pathol. Microbiol. *41*, 137–138 (1974)

Krag, S.S., Robbins, P.W.: Sindbis envelope proteins as endogenous acceptors in reactions of guanosine diphosphate-[^{14}C]mannose with preparations of infected chicken embryo fibroblasts. J. Biol. Chem. *252*, 2621–2629 (1977)

Lachmi, B., Kääriäinen, L.: Sequential translation of nonstructural proteins in cells infected with a Semliki Forest virus mutant. Proc. Natl. Acad. Sci. USA *73*, 1936–1940 (1976)

Lachmi, B., Kääriäinen, L.: Control of protein synthesis in Semliki Forest virus infected cells. J. Virol. *22*, 142–149 (1977)

Lachmi, B., Glanville, N., Keränen, S., Kääriäinen, L.: Tryptic peptide analysis of nonstructural and structural precursor proteins from Semliki Forest virus mutant-infected cells. J. Virol. *16*, 1615–1629 (1975)

Laemmli, U.K.: Cleavage of structural proteins during the assembly of the head of bacteriophage T4. Nature (London) *227*, 680–685 (1970)

Lagwinska, E., Stewart, C.C., Adles, C., Schlesinger, S.: Replication of lactic dehydrogenase virus and Sindbis virus in mouse peritoneal macrophages: Induction of interferon and phenotypic mixing. Virology *65*, 204–214 (1975)

Laine, R., Kettunen, M.-L., Gahmberg, C.G., Kääriäinen, L., Renkonen, O.: Fatty chains of different lipid classes of Semliki Forest virus and host cell membranes. J. Virol. *10*, 433–438 (1972)

Laine, R., Söderlund, H., Renkonen, O.: Chemical composition of Semliki Forest virus. Intervirology *1*, 110–118 (1973)

Lascano, E.F., Berria, M.I., Oro, J.G.B.: Morphogenesis of Aura virus. J. Virol. *4*, 271–282 (1969)

Lawrence, C., Thach, R.E.: Encephalomyocarditis virus infection of mouse plasmacytoma cells. I. Inhibition of cellular protein synthesis. J. Virol. *14*, 598–610 (1974)

Leavitt, R., Schlesinger, S., Kornfeld, S.: Tunicamycin inhibits glycosylation and multiplication of Sindbis and vesicular stomatitis viruses. J. Virol. *21*, 375–385 (1977)

Lenard, J., Compans, R.W.: The membrane structure of lipid containing viruses. Biochim. Biophys. Acta *344*, 51–94 (1974)

Levin, J.G., Friedman, R.M.: Analysis of arbovirus ribonucleic acid forms by polyacrylamide gel electrophoresis. J. Virol. *7*, 504–514 (1971)

Levin, J., Ramseur, J.M., Grimley, P.M.: Host effect on arbovirus replication: Appearance of defective interfering particles in murine cells. J. Virol. *12*, 1401–1406 (1973)

Lubiniecki, A.S.: Replication of eastern equine encephalitis viruses (New Jersey and Louisiana strains and the Ets-4 mutant) in rabbit kidney cells. Arch. Virol. *47*, 21–29 (1975)

Lubiniecki, A.S., Henry, C.J.: Autoradiographic localization of RNA synthesis directed by arboviruses in the cytoplasm of infected BHK-21 cells. Proc. Soc. Exp. Biol. Med. *145*, 1165–1169 (1974)

Lust, G.: Alterations of protein synthesis in arbovirus-infected L cells. J. Bacteriol. *91*, 1612–1617 (1966)

Luukkonen, A., Gahmberg, C.G., Renkonen, O.: Surface labeling of Semliki Forest virus glycoproteins using galactose-oxidase. Virology *76*, 55–59 (1977a)

Luukkonen, A., Bonsdorff, C.-H. v., Renkonen, O.: Characterization of Semliki Forest virus grown in mosquito cells. Comparison with the virus from hamster cells. Virology *78*, 331–335 (1977 b)

Luukkonen, A., Kääriäinen, L., Renkonen, O.: Phospholipids of Semliki Forest virus grown in cultured mosquito cells. Biochim. Biophys. Acta *450*, 109–120 (1976)

Mantani, M., Kato, S.: Relationship between virus multiplication and culture cells synchronized by excess thymidine treatment. V. Effect of chikungunya virus infection upon the cell cycle. Biken J. *18*, 15–23 (1975)

Margotat, A., Laplane, J., Pisano, M.-R., Nicoli, J.: Polyuridylic sequences and negative strands of Sindbis virus-specific RNA: Study by affinity chromatography on poly (A)-Sepharose columns. Ann. Microbiol. (Paris) *127 B*, 242–256 (1976)

Marker, S.C., Connelly, D., Jahrling, P.B.: Receptor interaction between eastern equine encephalitis virus and chicken embryo fibroblasts. J. Virol. *21*, 981–985 (1977)

Martin, B.A.B., Burke, D.C.: The replication of Semliki Forest virus. J. Gen. Virol. *24*, 45–66 (1974)

Martin, E.M.: Studies on the RNA polymerase of some temperature-sensitive mutants of Semliki Forest virus. Virology *39*, 107–117 (1969)

Martin, E.M., Sonnabend, J.A.: Ribonucleic acid polymerase catalyzing synthesis of double-stranded arbovirus ribonucleic acid. J. Virol. *1*, 97–109 (1967)

Martire, G., Bonatti, S., Aliperti, G., DeGiuli, C., Cancedda, R.: Free and membrane-bound polyribosomes in BHK cells infected with Sindbis virus. J. Virol. *21*, 610–618 (1977)

Mattila, K., Luukkonen, A., Renkonen, O.: Protein-bound oligosaccharides of Semliki Forest virus. Biochim. Biophys. Acta *419*, 435–444 (1976)

McCarthy, M., Harrison, S.C.: Glycosidase susceptibility: a probe for the distribution of glycoprotein oligosaccharides in Sindbis virus. J. Virol. *23*, 61–73 (1977)

McGee-Russel, S.M., Gosztonyi, G.: Assembly of Semliki Forest virus in brain. Nature (London) *214*, 1204–1206 (1967)

Michel, M.R., Gomatos, P.J.: Semliki Forest virus-specific RNAs synthesized in vitro by enzyme from infected BHK cells. J. Virol. *11*, 900–914 (1973)

Mims, C.A., Day, M.F., Marshall, I.D.: Cytopathic effect of Semliki Forest virus in the mosquito Aedes aegypti. Am. J. Trop. Med. Hyg. *15*, 775–784 (1966)

Mooney, J.J., Dalrymple, J.M., Alving, C.R., Russell, P.K.: Interaction of Sindbis virus with liposomal model membranes. J. Virol. *15*, 225–231 (1975)

Moore, N.F., Barenholz, Y., Wagner, R.R.: Microviscosity of togavirus membranes studied by fluorescence depolarization: Influence of envelope proteins and the host cell. J. Virol. *19*, 126–135 (1976)

Morgan, C., Howe, C.: Structure and development of viruses as observed in the electron microscope. IX. Entry of parainfluenza I (Sendai) virus. J. Virol. *2*, 1122–1132 (1968)

Morgan, C., Howe, C., Rose, H.N.: Structure of viruses as observed in the electron microscope. V. Western equine encephalomyelitis virus. J. Exp. Med. *113*, 219–234 (1961)

Morser, M.J., Burke, D.C.: Cleavage of virus-specific polypeptides in cells infected with Semliki Forest virus. J. Gen. Virol. *23*, 395–409 (1974)

Morser, M.J., Kennedy, S.I.T., Burke, D.C.: Virus-specific polypeptides in cells infected with Semliki Forest virus. J. Gen. Virol. *21*, 19–29 (1973)

Mowshowitz, D.: Identification of polysomal RNA in BHK cells infected by Sindbis virus. J. Virol. *11*, 535–543 (1973)

Murphy, F.A., Harrison, A.K., Collin, W.K.: The role of extraneural arbovirus infection in the pathogenesis of encephalitis. An electron microscopic study of Semliki Forest virus infection in mice. Lab. Invest. *22*, 318–328 (1970)

Mussgay, M., Rott, R.: Studies on the structure of a hemagglutinin component of a group A arbovirus (Sindbis). Virology *23*, 573–581 (1964)

Mussgay, M., Weibel, J.: Electron microscopic demonstration of purified Venezuelan equine encephalitis virus. Virology *19*, 109–112 (1963)

Mussgay, M., Enzmann, P.J., Horst, J.: Influence of an arbovirus infection (Sindbis virus) on the protein and ribonucleic acid synthesis of cultivated chick embryo cells. Arch. Ges. Virusforsch. *31*, 81–92 (1970)

Mussgay, M., Enzmann, P.J., Weiland, E., Horzinek, M.C.: Growth cycle of arboviruses in vertebrate and arthropod cells. Prog. Med. Virol. *19*, 258–323 (1975)

Mussgay, M., Weiland, E., Strohmaier, K., Ueberschär, S., Enzmann, P.J.: Properties of components obtained by treatment of Semliki Forest virus with Tween 80 and Tri(n-butyl)phosphate. J. Gen. Virol. *19*, 89–101 (1973)

Negro-Ponzi, A.: Sullàdsorbimento del virus Sindbis su monostrati di cellule di embrione di pollo. Ig. Mod. *60*, 944–949 (1967)

Neville, D.M.: Molecular weight determination of protein-dodecyl sulfate complexes by electrophoresis in a discontinuous buffer system. J. Biol. Chem. *246*, 6328–6334 (1971)

Nicoli, J.: Les recepteurs erythrocytaires des arbovirus. Les lipides globulaires. Ann. Inst. Pasteur (Paris) *109*, 472–478 (1965)

Oram, J.D., Ellwood, D.C., Appleyard, G., Stanley, J.L.: Agglutination of an arbovirus by concanavaline A. Nature New Biol. *233*, 50–51 (1971)

Osterrieth, P.M.: Semliki Forest virus inactivation with selective destruction of the haemagglutinating activity by caseinase C and of infectivity by deoxycholate. Life Sci. *4*, 1227–1231 (1965)

Osterrieth, P.M.: Recherches sur la structure et la fonction des couches externes du virus de la foret de la semliki: Modifications de celles-ci par l'action de la caseinase C. Mem. Soc. R. Sci. Liege Collect. 8 *16*, 11–125 (1968)

Osterrieth, P.M., Calberg-Bacq, C.M.: Changes in morphology, infectivity and haemagglutinating activity of Semliki Forest virus produced by the treatment with caseinase C from Streptomyces albus G. J. Gen. Microbiol. *43*, 19–30 (1966)

Palade, G.: Intracellular aspects of the process of protein synthesis. Science *189*, 347–358 (1975)

Pathak, S., Webb, H.E.: Possible mechanisms for the transport of Semliki Forest virus into and within mouse brain: An electronmicroscopic study. J. Neurol. Sci. *23*, 175–184 (1974)

Pathak, S., Webb, H.E., Oaten, S.W., Bateman, S.: An electronmicroscopic study of the development of virulent and avirulent strains of Semliki Forest virus in mouse brain. J. Neurol. Sci. *28*, 289–300 (1976)

Pedersen, C.E., Eddy, G.A.: Separation, isolation, and immunological studies of the structural proteins of Venezuelan equine encephalomyelitis virus. J. Virol. *14*, 740–744 (1974)

Pedersen, C.E., Eddy, G.A.: Comparative analyses of members of the Venezuelan equine encephalomyelitis virus complex. Am. J. Epidemiol. *101*, 245–252 (1975)

Pedersen, C.E., Marker, S.C., Eddy, G.A.: Comparative electrophoretic studies on the structural proteins of Selected group A arboviruses. Virology *60*, 312–314 (1974)

Peleg, J.: Inapparent persistent virus infection in continuously grown Aedes aegypti mosquito cells. J. Gen. Virol. *5*, 463–471 (1969)

Peleg, J.: Studies on the behavior of arboviruses in an Aedes aegypti mosquito cell line (Peleg). Ges. Virusforsch. *37*, 54–61 (1972)

Pesonen, M., Renkonen, O.: Serum glycoprotein-type sequence of monosaccharides in membrane glycoproteins of Semliki Forest virus. Biochim. Biophys. Acta *455*, 510–525 (1976)

Pesonen, M. Renkonen, O.: Structure of the complex carbohydrate chains in the membrane glycoproteins of Semliki Forest virus. Biochem. Trans. *5*, 120 (1977)

Pfefferkorn, E.R.: Comparative aspects of the replication of alphaviruses and poliovirus. Med. Biol. *53*, 337–341 (1975)

Pfefferkorn, E.R., Boyle, M.K.: Selective inhibition of the synthesis of Sindbis virion proteins by an inhibitor of chymotrypsin. J. Virol. *9*, 187–188 (1972)

Pfefferkorn, E.R, Burge, B.W.: Genetics and biochemistry of arbovirus temperature-sensitive mutants. In: The Molecular Biology of Viruses. Colter, J.S., Paranchych, W. (Eds.) New York: Academic Press, 1967, pp. 403–426

Pfefferkorn, E.R., Burge, B.W.: Morphogenic defects in the growth of ts-mutants of Sindbis virus. Perspect. Virol. *8*, 1–14 (1969)

Pfefferkorn, E.R., Clifford, R.L.: The origin of the protein of Sindbis virus. Virology *23*, 217–223 (1964)

Pfefferkorn, E.R., Hunter, H.S.: Purification and partial chemical analysis of Sindbis virus. Virology *20*, 433–445 (1963a)

Pfefferkorn, E.R., Hunter, H.S.: The source of the ribonucleic acid and phospholipid of Sindbis virus. Virology *20*, 446–456 (1963b)

Pfefferkorn, E.R., Shapiro, D.: Reproduction of togaviruses. In: Comprehensive Virology, Vol. 2. Fraenkel-Conrat, H., Wagner, R.R. (Eds.) New York and London: Plenum Press, 1974, pp. 171–230

Pfefferkorn, E.R., Burge, B.W., Coady, H.M.: Intracellular conversion of the RNA of Sindbis virus to a double-stranded form. Virology *33*, 239–249 (1967)

Pierce, J.S., Strauss, E.G., Strauss, J.H.: Effect of ionic strength on the binding of Sindbis virus to chick cells. J. Virol. *13*, 1030–1036 (1974)

Pinter, A., Compans, R.W.: Sulfated components of enveloped viruses. J. Virol. *16*, 859–866 (1975)

Pisano, M.R, Poiree, J.C., Lanuc, L., Euret, M., Nicoli, J.: The length of polyadenylic acid tracts of Sindbis RNA: The possible existence of two synthetic mechanisms. C.R. Acad. Sci. (Paris) *280*, 1023–1026 (1975)

Poiree, J.-C., Bonnet, J., Negre, G., Nicoli, J.: Polyribosomes et acides ribonucleiques associes aux polyribosomes dans les cellules infectees par un togavirus, le virus Sindbis. Ann. Microbiol. (Paris) *124*, 371–385 (1973)

Poiree, J.-C., Bonneau, H.-P., Nicoli, J.: Les acides ribonucleiques monocatenaires d'un Arbovirus de groupe A: Le virus Sindbis. C.R. Hebd. Seances Acad. Sci. Ser D Sci. Nat. (Paris) *274*, 741–744 (1972)

Precious, S.W., Webb, H.E., Bowen, E.T.W.: Isolation and persistence of Chikungunya virus in cultures of mouse brain cells. J. Gen. Virol. *23*, 271–279 (1974)

Quersin-Thiry, L.: Interaction between cellular extracts and animal viruses. I. Kinetic studies and some notes on the specificity of the interaction. Acta Virol. *5*, 141–152 (1961)

Quersin-Thiry, L., Nihoul, E.: Interaction between cellular extracts and animal viruses. II. Evidence for the presence of different inactivators corresponding to different viruses. Acta Virol. *5*, 283–293 (1961)

Raghow, R.S., Davey, M.W., Dalgarno, L.: The growth of Semliki Forest virus in cultured mosquito cells: Ultrastructural observations. Brief report. Arch. Ges. Virusforsch. *43*, 165–168 (1973)

Ranki, M., Kääriäinen, L.: The effect of canavanine on Semliki Forest virus RNA synthesis. Ann. Med. Exp. Biol. Fenn. *48*, 238–245 (1970)

Ranki, M., Kääriäinen, L., Renkonen, O.: Semliki Forest virus glycoproteins and canavanine. Acta Pathol. Microbiol. Scand. Sect. B *80*, 760–768 (1972)

Rasilo, M.-L., Renkonen, O.: Hydrazinolysis of Semliki Forest virus glycopeptides. Manuscript in preparation. (1978)

Ravid, Z., Goldblum, N.: Analysis of antigenic activities of enzymatically digested Sindbis virus envelopes. Intervirology *2*, 152–159 (1973/74)

Renkonen, O., Gahmberg, C.G., Simons, K., Kääriäinen, L.: The lipids of the plasma membrane and endoplasmic reticulum from cultured baby hamster kidney cells (BHK21). Biochim. Biophys. Acta *255*, 66–78 (1972a)

Renkonen, O., Kääriäinen, L., Gahmberg, C.G., Simons, K.: Lipids of Semliki Forest virus and host cell membranes. In: Current Trends in the Biochemistry of Lipids. Ganguly, J., Smellie, R.M. S. (Eds.) New York: Academic Press, 1972b, pp. 407–422

Renkonen, O., Kääriäinen, L., Simons, K., Gahmberg, C.G.: The lipid class composition of Semliki Forest virus and of plasma membranes of the host cells. Virology *46*, 318–326 (1971)

Renkonen, O., Luukkonen, A., Brotherus, J., Kääriäinen, L.: Composition and turnover of membrane lipids in Semliki Forest virus and in host cells. In: Control of Proliferation in Animal Cells. Clarkson, B., Baserga, R. (Eds.) New York: Cold Spring Harbor Laboratory 1974, pp. 495–504

Renkonen, O., Pesonen, M., Mattila, K.: Oligosaccharides of the membrane glycoproteins of Semliki Forest virus. In: Structure of Biological Membranes. Abrahamson, L., Pascher, I. (Eds.) New York: Plenum Press, 1976, pp. 409–423

Renz, D., Brown, D.T.: Characteristics of Sindbis virus temperature-sensitive mutants in cultured BHK-21 and Aedes albopictus (mosquito) cells. J. Virol. *19*, 775–781 (1976)

Richardson, C.D., Vance, D.E.: Biochemical evidence that Semliki Forest virus obtains its envelope from the plasma membrane of the host cell. J. Biol. Chem. *251*, 5544–5550 (1976)

Rubin, H., Baluda, M., Hotchin, J.E.: The maturation of western equine encephalomyelitis virus and its release from chick embryo cell suspensions. J. Exp. Med. *101*, 205–212 (1955)

Rueckert, R.R.: On the structure and morphogenesis of picornaviruses. In: Comprehensive Virology, Vol. 6 Fraenkel-Conrat, H., Wagner, R.R. (Eds.) New York: Plenum Press, 1976, pp. 131–213

Saborio, J.L., Pong, S.S., Koch, G.: Selective and reversible inhibition of protein synthesis in mammalian cells. J. Mol. Biol. *85*, 195–211 (1974)

Salminen, A.: Chemistry of nonspecific inhibitors of hemagglutination by arthropod-borne viruses. Virology *16*, 201–203 (1962)

Saraste, J., Kääriäinen, L., Söderlund, H., Keränen, S.: RNA synthesis directed by a temperature-sensitive mutant of Semliki Forest virus. J. Gen. Virol. *37*, 399–406 (1977)

Sarver, N., Stollar, V.: Sindbis virus-induced cytopathic effect in clones of Aedes albopictus (Singh) cells. Virology *80*, 390–400 (1977)

Sawicki, D.L., Gomatos, P.J.: Replication of Semliki Forest virus: Polyadenylate in plus-stranded RNA and polyuridylate in minus-stranded RNA. J. Virol. *20*, 446–464 (1976)

Sawicki, D., Kääriäinen, L., Lambeck, C., Gomatos, P.J.: Mechanism for control of synthesis of Semliki Forest virus 26S and 42S RNA. J. Virol. *25*, 19–27 (1978)

Scheele, C.M., Pfefferkorn, E.R.: Kinetics of incorporation of structural proteins into Sindbis virions. J. Virol. *3*, 369–375 (1969a)

Scheele, C.M., Pfefferkorn, E.R.: Inhibition of interjacent ribonucleic acid (26S) synthesis in cells infected by Sindbis virus. J. Virol. *4*, 117–122 (1969b)

Scheele, C.M., Pfefferkorn, E.R.: Virus-specific proteins synthesized in cells infected with RNA$^+$ temperature-sensitive mutants of Sindbis virus. J. Virol. *5*, 329–337 (1970)

Schlesinger, M.J., Schlesinger, S.: Large-molecular-weight precursors of Sindbis virus proteins. J. Virol. *11*, 1013–1016 (1973)

Schlesinger, M.J., Schlesinger, S., Burge, B.W.: Identification of a second glycoprotein in Sindbis virus. Virology *47*, 539–541 (1972)

Schlesinger, R.W.: Sindbis virus replication in vertebrate and mosquito cells: an interpretation. Med. Biol. *53*, 295–301 (1975)

Schlesinger, S., Schlesinger, M.J.: Formation of Sindbis virus proteins: Identification of a precursor for one of the envelope proteins. J. Virol. *10*, 925–932 (1972)

Schlesinger, S., Gottlieb, C., Feil, P., Gelb, N., Kornfeld, S.: Growth of enveloped RNA viruses in a line of chinese hamster ovary cells with deficient N-acetyl-glucosaminyltransferase activity. J. Virol. *17*, 239–246 (1976)

Schlesinger, S., Schlesinger, M., Burge, B.W.: Defective virus particles from Sindbis virus. Virology *48*, 615–617 (1972)

Schlesinger, S., Weiss, B., Dohner, D.: Defective particles in alphavirus infections. Med. Biol. *53*, 372–379 (1975)

Schlesinger, S., Weiss, B., Goran, D., Schlesinger, M., Cancedda, R.: Formation of RNA and protein in cells infected with standard and defective Sindbis virus. Med. Microbiol. Immunol. *160*, 311–329 (1974)

Schmidt, M.F.G., Schwarz, R.T., Scholtissek, C.: Interference of nucleoside diphosphate derivatives of 2-deoxy-D-glucose with the glycosylation of virus-specific glycoproteins in vivo. Eur. J. Biochem. *70*, 55–62 (1976)

Scholtissek, C.: Inhibition of the multiplication of enveloped viruses by glucose derivatives. Curr. Top. Microbiol. Immunol. *70*, 101–124 (1975)

Scholtissek, C., Kaluza, G.: Interference with the glycosylation of Semliki Forest virus proteins. Med. Biol. *53*, 357–364 (1975)

Scholtissek, C., Kaluza, G., Rott, R.: Stability and precursor relationships of virus RNA. J. Gen. Virol. *17*, 213–219 (1972)

Scholtissek, C., Kaluza, G., Schmidt, M., Rott, R.: Influence of sugar derivatives on glycoprotein synthesis of enveloped viruses. In: Negative strand Viruses, Vol. 2, Mahy, B.W.J., Barry, R.D. (Eds.) New York: Academic Press, 1975 pp. 669–683

Schwarz, R.T., Rohrschneider, J.M., Schmidt, M.F.G.: Suppression of glycoprotein formation of Semliki Forest, influenza, and avian sarcoma virus by tunicamycin. J. Virol. *19*, 782–791 (1976)

Schwoebel, W., Ahl, R.: Persistence of Sindbis virus in BHK-21 cell cultures. Arch. Ges. Virusforsch. *38*, 1–10 (1972)

Schwoebel, W., Speiser, C., Warnke, M.K.: Changes of the cell type of BHK-21 lines by persistent Sindbis virus. Zentralbl. Bakteriol. (orig. A) *231*, 42–46 (1975)

Scupman, R.K., Jones, K.J., Sagik, B.P., Bose, H.R.: Virus-directed post-translational cleavage in Sindbis virus-infected cells. J. Virol. *22*, 568–571 (1977)

Seamer, J.: Limited infection of mouse brain cell cultures with Semliki Forest virus. Br. J. Exp. Pathol. *55*, 606–614 (1974)

Sefton, B.M.: Virus-dependent glycosylation. J. Virol. *17*, 85–93 (1976)

Sefton, B.M.: Immediate glycosylation of Sindbis virus membrane proteins. Cell *10*, 659–668 (1977)

Sefton, B.M., Burge, B.W.: Biosynthesis of the Sindbis virus carbohydrates. J. Virol. *12*, 1366–1374 (1973)

Sefton, B.M., Gaffney, B.J.: Effect of the viral proteins on the fluidity of the membrane lipids in Sindbis virus. J. Mol. Biol. *90*, 343–358 (1974)

Sefton, B.M., Keegstra, K.: Glycoproteins of Sindbis virus: Preliminary characterization of the oligosaccharides. J. Virol. *14*, 522–530 (1974)

Sefton, B.M., Wickus, G.G., Burge, B.W.: Enzymatic iodination of Sindbis virus proteins. J. Virol. *11*, 730–735 (1973)

Segal, S., Streevalsan, T.: Sindbis virus replicative intermediates: Purification and characterization. Virology *59*, 428–442 (1974)

Shenk, T.E., Stollar, V.: Viral RNA species in BHK-cells infected with Sindbis virus serially passaged at high multiplicity of infection. Biochem. Biophys. Res. Commun *49*, 60–67 (1972)

Shenk, T.E., Stollar, V.: Defective-interfering particles of Sindbis virus: I. Isolation and some chemical and biological properties. Virology *53*, 162–173 (1973a)

Shenk, T.E., Stollar, V.: Defective-interfering particles of Sindbis virus. II. Homologous interference. Virology *55*, 530–534 (1973b)

Shenk, T.E., Koshelnyk, K.A., Stollar, V.: Temperature-sensitive virus from Aedes albopictus cells chronically infected with Sindbis virus. J. Virol. *13*, 439–447 (1974)

Simizu, B., Wagatsuma, M., Oya, A., Kanaoka, F., Yamada, M.: Inhibition of cellular DNA synthesis in hamster kidney cells infected with western equine encephalitis virus. Arch. Virol. *51*, 251–261 (1976)

Simizu, B., Yamazaki, S., Suzuki, K., Terasima, T.: Gamma ray-induced small plaque mutants of western equine encephalitis virus. J. Virol. *12*, 1568–1578 (1973)

Simmons, D.T., Strauss, J.H.: Replication of Sindbis virus: I. Relative size and genetic content of 26 S and 49 S RNA. J. Mol. Biol. *71*, 599–613 (1972a)

Simmons, D.T., Strauss, J.H.: Replication of Sindbis virus: II. Multiple forms of double-stranded RNA isolated from infected cells. J. Mol. Biol. *71*, 615–631 (1972b)

Simmons, D.T., Strauss, J.H.: Replication of Sindbis virus: V. Polyribosomes and mRNA in infected cells. J. Virol. *14*, 552–559 (1974a)

Simmons, D.T., Strauss, J.H.: Translation of Sindbis virus 26 S RNA and 49 S RNA in lysates of rabbit reticulocytes. J. Mol. Biol. *86*, 397–409 (1974b)

Simons, K., Kääriäinen, L.: Characterization of the Semliki Forest virus core and envelope protein. Biochem. Biophys. Res. Commun. *38*, 981–988 (1970)

Simons, K., Garoff, H., Helenius, A., Kääriäinen, L., Renkonen, O.: Structure and assembly of virus membranes. In: Perspectives in Membrane Biology. Estrado, O., Gitler, C. (Eds.) New York: Academic Press, 1974, p. 45

Simons, K., Garoff, H., Helenius, A.: The glycoproteins of Semliki Forest virus membrane. In: Membrane Proteins and their Interactions with Lipids. Capaldi, R.A. (Ed.) New York: Marcel Dekker, 1977, pp. 59–86

Simons, K., Garoff, H., Helenius, A., Ziemiecki, A.: The structure and assembly of the membrane of Semliki Forest virus. In: Frontiers of Physicochemical Biology. Pullmann, B. (Ed.) New York: Academic Press, 1978 (in press)

Simons, K., Helenius, A., Garoff, H.: Solubilization of the membrane proteins from Semliki Forest virus. J. Mol. Biol. *80*, 119–133 (1973a)

Simons, K., Keränen, S., Kääriäinen, L.: Identification of a precursor for one of the Semliki Forest virus membrane proteins. FEBS Lett. *29*, 87–91 (1973b)

Simpson, R.W., Hauser, R.E.: Basic structure of group A arbovirus strains Middelburg, Sindbis, and Semliki Forest examined by negative staining. Virology *34,* 358–361 (1968)

Singh, K.R.P., Paul, S.D.: Susceptibility of Aedes albopictus and Aedes aegypti cell lines to infection by arbo and other viruses. Indian J. Med. Res. *56,* 815–819 (1968)

Sly, W.S., Lagwinska, E., Schlesinger, S.: Enveloped virus acquires membrane defect when passaged in fibroblasts from I-cell disease patients. Proc. Natl. Acad. Sci. USA *73,* 2443–2447 (1976)

Smith, A.E.: The initiation of protein synthesis directed by the RNA from encephalomyocarditis virus. Eur. J. Biochem. *33,* 301–313 (1973)

Smith, A.E., Wheeler, T., Glanville, N., Kääriäinen, L.: Translation of Semliki-Forest-virus 42-S RNA in a mouse cell-free system to give virus-coat proteins. Eur. J. Biochem. *49,* 101–110 (1974)

Smith, J.F., Brown, D.T.: Envelopment of Sindbis virus: synthesis and organization of proteins in cells infected with wild type and maturation defective mutants. J. Virol. *22,* 662–678 (1977)

Snyder, H.W., Streevalsan, T.: Proteins specified by Sindbis virus in HeLa cells. J. Virol. *13,* 541–544 (1974)

Söderlund, H.: Kinetics of formation of the Semliki Forest virus nucleocapsid. Intervirology *1,* 354–361 (1973)

Söderlund, H.: The post-translational processing of Semliki Forest virus structural polypeptides in puromycin treated cells. FEBS Lett. *63,* 56–58 (1976)

Söderlund, H., Kääriäinen, L.: Association of capsid protein with Semliki Forest virus messenger RNAs. Acta Path. Microbiol. Scand. Sect. B. *82,* 33–40 (1974)

Söderlund, H., Ulmanen, I.: Transient association of Semliki Forest virus capsid protein with ribosomes. J. Virol. *24,* 907–909 (1977)

Söderlund, H., Glanville, N., Kääriäinen, L.: Polysomal RNAs in Semliki Forest virus-infected cells. Intervirology *2,* 100–113 (1973/74)

Söderlund, H., Kääriäinen, L., Bonsdorff, C.-H. v., Weckström, P.: Properties of Semliki Forest virus nucleocapsid: II. An irreversible contaction by acid pH. Virology *47,* 753–760 (1972)

Söderlund, H., Kääriäinen, L., Bonsdorff, C.-H. v.: Properties of Semliki Forest virus nucleocapsid. Med. Biol. *53,* 412–417 (1975)

Sonnabend, J., Dalgarno, L., Friedman, R.M., Martin, E.M.: A possible replicative form of Semliki Forest virus RNA. Biochem. Biophys. Res. Commun. *17,* 455–460 (1964)

Sonnabend, J.A., Martin, E.M., Mécs, É.: Viral specific RNAs in infected cells. Nature (London) *213,* 365–367 (1967)

Spiro, R.G.: Glycoproteins. Adv. Protein Chem. *27,* 349–467 (1973)

Sprecher-Goldberger, S.: Differences between the structures of poliovirus and Sindbis virus infectious ribonucleic acids. Arch. Ges. Virusforsch. *20,* 225–234 (1967)

Sreevalsan, T.: Association of viral ribonucleic acid with cellular membranes in chick embryo cells infected with Sindbis virus. J. Virol. *6,* 438–444 (1970)

Sreevalsan, T., Allen, P.T.: Replication of western equine encephalomyelitis virus. II. Cytoplasmic structure involved in the synthesis and development of the virions. J. Virol. *2,* 1038–1046 (1968)

Sreevalsan, T., Lockart, R.Z.: Inhibition by puromycin of the initiation of synthesis of infectious RNA and virus by chicken embryo cells infected with western equine encephalomyelitis virus. Virology *24,* 91–96 (1964)

Sreevalsan, T., Lockart, R.Z.: Heterogenous RNA's occurring during the replication of western equine encephalomyelitis virus. Proc. Natl. Acad. Sci. USA *55,* 974–981 (1966)

Sreevalsan, T., Yin, F.H.: Sindbis virus-induced viral ribonucleic acid polymerase. J. Virol. *3,* 599–604 (1969)

Sreevalsan, T., Lockart, R.Z., Dodson, M.L., Hartman, K.A.: Replication of western equine encephalomyelitis virus: I. Some chemical and physical characteristics of viral ribonucleic acid. J. Virol. *2,* 558–566 (1968)

Stern, R., Friedman, R.M.: Chromatography of arbovirus ribonucleic acid forms on columns of benzoylated-diethylaminoethyl cellulose. J. Virol. *4,* 356–364 (1969)

Stevens, T.M.: Arbovirus replication in mosquito cell lines (Singh) grown in monolayer or suspension culture. Proc. Soc. Exp. Biol. Med. *134,* 356–361 (1970)

Stoffel, W., Sorgo, W.: Asymmetry of the lipid-bilayer of Sindbis virus. Chem. Phys. Lipids *17*, 324–335 (1976)

Stollar, B.D., Stollar, V.: Immunofluorescent demonstration of double-stranded RNA in the cytoplasm of Sindbis virus-infected cells. Virology *42*, 276–280 (1970)

Stollar, V., Peleg, J., Shenk, T.E.: Temperature sensitivity of a Sindbis virus mutant isolated from persistently infected Aedes aegypti cell culture. Intervirology *2*, 337–344 (1973/74)

Stollar, V., Skenk, T.E., Stollar, B.D.: Double-stranded RNA in hamster, chick and mosquito cells infected with Sindbis virus. Virology *47*, 122–132 (1972)

Stollar, V., Stollar, B.D., Koo, R., Harrap, K.A., Schlesinger, R.W.: Sialic acid contents of Sindbis virus from vertebrate and mosquito cells. Equivalence of biological and immunological viral properties. Virology *69*, 104–115 (1976)

Strauss, E.G., Birdwell, C.R., Lenches, E.M., Staples, S.E., Strauss, J.H.: Mutants of Sindbis virus. II. Characterization of a maturation defective mutant, ts-103. Virology *82*, 122–149 (1977)

Strauss, E.G., Lenches, E.M., Strauss, J.H.: Mutants of Sindbis virus: Isolation and partial characterization of 89 new temperature-sensitive mutants. Virology *74*, 154–168 (1976)

Strauss, J.H., Strauss, E.G.: Togaviruses. In: The Molecular Biology of Animal Viruses Nayak, D.P. (Ed.) New York: Marcel Dekker, 1977, pp. 111–166

Strauss, J.H., Burge, B.W., Darnell, J.E.: Sindbis virus infection of chick and hamster cells: synthesis of virus-specific proteins. Virology *37*, 367–376 (1969)

Strauss, J.H., Burge, B.W., Darnell, J.E.: Carbohydrate content of the membrane protein of Sindbis virus. J. Mol. Biol. *47*, 437–448 (1970)

Strauss, J.H., Burge, B.W., Pfefferkorn, E.R., Darnell, J.E.: Identification of the membrane protein and "core" protein of Sindbis virus. Proc. Natl. Acad. Sci. USA *59*, 533–537 (1968)

Struck, D.K., Lennarz, W.Y.: Evidence for the participation of saccharide-lipids in the synthesis of the oligosaccharide chain of ovalbumin. J. Biol. Chem. *252*, 1007–1013 (1977)

Takatsuki, A., Kohno, K., Tamura, G.: Inhibition of biosynthesis of polyisoprenol sugars in chick embryo microsomes by tunicamycin. Agric. Biol. Chem. *39*, 2089–2091 (1975)

Takehara, M.: Comparative studies on nucleic acid synthesis and virus-induced RNA polymerase activity in mammalian cells infected with certain arboviruses. Arch. Ges. Virusforsch. *34*, 266–277 (1971)

Takehara, M.: Inhibition of nuclear protein synthesis in BHK 21 cells infected with arboviruses. Arch. Ges. Virusforsch. *39*, 163–171 (1972)

Tan, K.B.: Electron microscopy of cells infected with Semliki Forest virus temperature-sensitive mutants: Correlation of ultrastructural and physiological observations. J. Virol. *5*, 632–638 (1970)

Tan, K.B.: Comparative study of the protein kinase associated with animal viruses. Virology *64*, 566–570 (1975)

Tan, K.B., Sokol, F.: Virion-bound protein kinase in Semliki Forest and Sindbis viruses. J. Virol. *13*, 1245–1253 (1974)

Tan, K.B., Sambrook, J.F., Bellet, A.J.P.: Semliki Forest temperature-sensitive mutants: isolation and characterization. Virology *38*, 427–439 (1969)

Taylor, J.: Studies on the mechanism of action of interferon. I. Interferon action and RNA synthesis in chick embryo fibroblasts infected with Semliki Forest virus. Virology *25*, 340–349 (1965)

Tkacz, J.S., Lampen, O.: Tunicamycin inhibition of polyisoprenol N-acetylglucosaminyl pyrophosphate formation in calf liver microsomes. Biochem. Biophys. Res. Commun. *65*, 248–257 (1975)

Tuomi, K., Kääriäinen, L., Söderlund, H.: Quantitation of Semliki Forest virus RNAs in infected cells using ^{32}P equilibrium labelling. Nucleis Acids. Res. *2*, 555–565 (1975)

Ueba, N., Kimura, T.: Polykaryocytosis induced by certain arboviruses in monolayers of BHK-21-528 cells. J. Gen. Virol. *34*, 369–373 (1977)

Ulmanen, I., Söderlund, H., Kääriäinen, L.: Semliki Forest virus capsid protein associates with the 60S ribosomal subunit in infected cells. J. Virol. *20*, 203–210 (1976)

Uryayev, L.V., Zhdanov, V.M., Yershov, F.I., Bykovsky, A.F.: Characteristics of Venezuelan equine encephalomyelitis virus ribonucleoprotein. Arch. Ges. Virusforsch. *33*, 281–287 (1971)

Uterman, G., Simons, K.: Studies on the amphipathic nature of the membrane proteins in Semliki Forest virus. J. Mol. Biol. *85*, 569–587 (1974)

Veckenstedt, A., Wagner, M.: Virological and immunofluorescent studies on the multiplication of Sindbis virus in L cells. Arch. Ges. Virusforsch. *42*, 144–153 (1973)

Villa-Komaroff, L., Guttman, N., Baltimore, D., Lodish, H.F.: Complete translation of poliovirus RNA in a eukaryotic cell-free system. Proc. Natl. Acad. Sci. USA *72*, 4157–4161 (1975)

Virtanen, I., Wartiovaara, J.: Virus-induced cytoplasmic membrane structures associated with Semliki Forest virus infection studied by the freeze-etching method. J. Virol. *13*, 222–225 (1974)

Wagner, B., Veckenstedt, A., Wagner, M.: Crystalline arrays of nucleocapsids in Sindbis virus-infected L cells. Intervirology *5*, 103–107 (1975)

Waite, M.R.F.: Protein synthesis directed by an RNA⁻ temperature-sensitive mutant of Sindbis virus. J. Virol. *11*, 198–206 (1973)

Waite, M.R.F., Brown, D.T., Pfefferkorn, E.R.: Inhibition of Sindbis virus release by media of low ionic strength: An electron microscope study. J. Virol. *10*, 537–544 (1972)

Waite, M.R.F., Lubin, M., Jones, K., Bose, H.R.: Phosphorylated proteins of Sindbis virus. J. Virol. *13*, 244–246 (1974)

Wecker, E.: Characteristics of an infectious nucleic acid fraction from chicken embryos infected with encephalitis virus. I. Physical and chemical characteristics. Zeitschr. Naturforsch. *14b*, 370–378 (1959a)

Wecker, E.: The extraction of infectious virus nucleic acid with hot phenol. Virology *7*, 241–243 (1959b)

Wecker, E.: Effect of puromycin on the replication of western equine encephalitis and poliomyelitis viruses. Nature (Lond.) *197*, 1277–1279 (1963)

Wecker, E., Schäfer, W.: An infectious component of ribonucleic acid character from the virus of the American equine encephalomyelitis (eastern type). Z. Naturforsch. *12b*, 415–417 (1957)

Wecker, E., Hummeler, K., Goetz, O.: Relationship between viral RNA and viral protein synthesis. Virology *17*, 110–117 (1962)

Weiss, B., Schlesinger, S.: Defective interfering passages of Sindbis virus: Chemical composition, biological activity and mode of interference. J. Virol. *12*, 862–871 (1973)

Weiss, B., Goran, D., Cancedda, R., Schlesinger, S.: Defective interfering passages of Sindbis virus: Nature of the intracellular defective viral RNA. J. Virol. *14*, 1189–1198 (1974)

Wengler, G., Wengler, G.: Studies on the polyribosome-associated RNA in BHK21 cells infected with Semliki Forest virus. Virology *59*, 21–35 (1974)

Wengler, G., Wengler, G.: Comparative studies on polyribosomal, nonpolyribosome-associated and viral 42S RNA from BHK 21 cells infected with Semliki Forest virus. Virology *65*, 601–605 (1975a)

Wengler, G., Wengler, G.: Studies on the synthesis of viral RNA-polymerase-template complexes in BHK 21 cells infected with Semliki Forest virus. Virology *66*, 322–326 (1975b)

Wengler, G., Wengler, G.: Localization of the 26-S RNA sequence on the viral genome 42-S RNA isolated from SFV-infected cells. Virology *73*, 190–199 (1976a)

Wengler, G., Wengler, G.: Protein synthesis in BHK 21 cells infected with Semliki Forest virus. J. Virol. *17*, 10–19 (1976b)

Wengler, G., Beato, M., Hackemack, B.-A.: Translation of 26 S virus-specific RNA from Semliki Forest virus-infected cells in vitro. Virology *61*, 120–128 (1974)

Wengler, G., Wengler, G., Filipe, A.R.: A study of nucleotide sequence homology between the nucleic acids of different alpha-viruses. Virology *78*, 124–134 (1977)

Wengler, G., Wengler, G., Wahn, K.: Isolation and characterization of double-stranded RNA containing infectious viral genome RNA from cells infected with Semliki Forest virus. Arch. Virol. *50*, 45–53 (1976)

Whitfield, S.G., Murphy, F.A., Sudia, W.D.: Eastern equine encephalomyelitis virus: An electron microscopic study of Aedes triseriatus (Say) salivary gland infection. Virology *43*, 110–122 (1971)

Wildy, P. (Ed.): International committee on nomenclature of viruses, classification and nomenclature of viruses: first report 1966–1970. In: Monographs in Virology, Basel: S. Karger, 1971, Vol. 5 pp. 52–54

Wiley, D., Bonsdorff, C.-H. v.: Three dimensional crystals of the membraned Semliki Forest virus. J. Mol. Biol. *120*, 375–379 (1978)

Wirth, D.F., Katz, F., Small, B., Lodish, H.: How a single-stranded Sindbis virus mRNA directs the synthesis of one soluble protein and two integral membrane glycoproteins. Cell *10*, 253–263 (1977)

Wittek, R., Koblet, H., Kohler, U., Wyler, R.: Poly-Adenylsäure-Abschnitte in Ribonukleinsäuren von Semliki-Forest-Virus. Pathol. Microbiol. *41*, 139–140 (1974)

Wittek, R., Koblet, H., Menna, A., Wyler, R.: The effect of cordycepin on the multiplication of Semliki Forest virus and on polyadenylation of viral RNA. Arch. Virol. *54*, 95–106 (1977)

Yin, F.H.: Temperature-sensitive behavior of the hemagglutinin in a temperature-sensitive mutant virion of Sindbis virus. J. Virol. *4*, 547–548 (1969)

Yoshinaka, Y., Hotta, S.: Infectivity of virus-specific RNAs of a group A arbovirus, Chikungunya. Virology *45*, 524–526 (1971)

Yoshinaka, Y., Hotta, S.: Chromatography of viral RNA forms produced during infection of a group A arbovirus, Chikungunya. Kobe J. Med. Sci. *18*, 153–168 (1972)

Zhdanov, V.M., Gaidamovich, S.Y., Melnikova, E.E., Krasnobayeva, Z.N.: Biophysical characteristics of Sindbis virion components. Acta Virol. *16*, 97–102 (1972)

Ziemiecki, A., Garoff, H.: Subunit structure of the Semliki Forest virus membrane proteins. J. Mol. Biol. in press (1978)

Initiation of DNA Synthesis by RNA

Kasper Zechel[1]

[1] Max-Planck-Institut für biophysikalische Chemie, Am Faßberg, D-3400 Göttingen-Nikolausberg

Abbreviations: Col E1 = colicinogenic factor E1 (plasmid coding for colicin E1); *E. coli* = *Escherichia coli*; Eco RI = restriction endonuclease from *E. coli* bearing the resistance transfer factor I (plasmid); Hpa II = restriction endonuclease from *Haemophilus parainfluenza;* iRNA = initiator RNA; RF I = closed, circular, double-stranded replicative form of DNA; RF II = circular, double-stranded replicative form of DNA having a discontinuity in one strand; SS = single-strand; SV 40 = simian virus 40.

I. Introduction

Up to the present one of the basic problems for the understanding of DNA replication is how the synthesis of DNA chains is initiated in vivo. So far none of the known DNA polymerases of prokaryotic or eukaryotic origin has been shown to initiate the synthesis of a DNA chain de novo. All DNA polymerases need a primer, which, in its simplest form, may be an oligoribo- or an oligodeoxyribonucleotide having a free 3'-hydroxyl end (*Karkas* and *Chargaff,* 1966). This primer terminus can be extended at its 3'-OH end by addition of new nucleotides as directed by the polynucleotide template to which the primer is linked through hydrogen bonds. Although the requirement for a primer has always been realized, earlier studies on DNA synthesis focused more on the mechanism of chain elongation than on the initiation process.

In order to explain how the necessary primer termini could become available inside a cell, several possibilities were considered. One was, for example, that a pool of small oligonucleotides may exist in the cells, consisting of either partial breakdown products of cellular nucleic acids or of specific primer oligonucleotides that would base-pair with the DNA and serve as primers. The basis for this hypothesis was provided by the observation that DNA synthesis by DNA polymerase I in vitro could be stimulated by heated extracts from *E. coli,* which were shown to contain fragmented DNA (*Goulian,* 1968). However, despite extensive search no such cellular primer oligonucleotides have been detected.

An alternative model calls for specific nucleases that would introduce scissions into a DNA strand at one or several specific site(s). In that way, 3'-OH ends could be created that would serve as primer terminus for extension by DNA polymerase. For this model, a number of nucleases have been discussed as possible initiators of DNA replication. The formation of the DNA protein relaxation complexes, which may be involved in the replication of the colicinogenic factors E1 and E2 DNAs, may require the action of a specific nuclease (*Blair* et al., 1971), and a model was proposed involving a hypothetical phage-coded "nickase" for the initiation of the replication of T4 DNA (*Kozinski,* 1968). The creation of a 3'-OH end for primer purposes by a specific nuclease seems to be exemplified in the initiation of the viral DNA synthesis of the bacteriophages φX174 and M13. A specific nuclease is coded for by the gene A of bacteriophage φX174 (*Francke* and *Ray,* 1972; *Henry* and *Knippers,* 1974; *Fujisawa* and *Hayashi,* 1976; *Ikeda* et al., 1976). It acts in vivo in cis fashion specifically on the replicative form DNA of bacteriophage φX174 and introduces a nick into the viral strand. The 3'-OH end of this nick serves as primer for a "rolling circle" type DNA synthesis. Since the gene A product is required for both the RF to RF replication and the subsequent single-stranded synthesis of bacteriophage φX174 DNA in vitro it appears that it may have additional functions (*Eisenberg* et al., 1977). The analogous protein for M13 reproduction is the gene 2 product of this phage (*Fidanián* and *Ray*, 1972; *Lin* and *Pratt*, 1972).

In vitro, many DNA polymerases have been shown to accept oligoribonucleotides as primers (*Wells* et al., 1972; *Chang* and *Bollum*, 1972; *Keller*, 1972; *Tamblyn* and *Wells*, 1975; *Plevani* and *Chang*, 1977; *Chargaff*, 1976). A number of experiments provided evidence that priming of DNA synthesis by RNA might also occur in vivo (*Brutlag* et al., 1971; *Crippa* and *Tocchini-Valentini*, 1971; *Ficq* and *Brachet*, 1971). This priming mechanism was attractive because the primer RNA could be synthesized by an RNA polymerase, since RNA polymerases, in contrast to DNA polymerases, are able to initiate ribonucleotide chains de novo on a duplex template (*Maitra* and *Hurwitz*, 1965). A brief transcriptional event therefore would be adequate to provide the system with a primer that in turn could be extended by a DNA polymerase. The question has often been raised as to why RNA primers may be of particular advantage over DNA primers. It appears plausible that the answers to this question is related to the ultra-high fidelity that is required for the copying of the DNA and the error-free preservation of the genetic information. As has been stressed in the context of the fidelity requirement in a review article by *Alberts* and *Sternglanz* (1977), the initiation of DNA chains bears the highest risk of introducing mistakes. An RNA primer is likely to be more easily recognized as a foreign piece of nucleic acid and erased than a DNA primer. With its removal any error that may have occurred at this step is automatically corrected. Since the formulation of the RNA primer hypothesis several years ago, considerable evidence has been accumulated that indeed RNA priming seems to be one major way of providing the necessary primers for DNA synthesis. To prove the participation of RNA, different approaches have been used in a variety of systems. The purpose of this review is to summarize these approaches and to discuss the significance of the results obtained for the assessment of the role of RNA priming of DNA synthesis. I am aware that this review covers only a very special aspect of DNA replication. Therefore, for a broader view of DNA synthesis and for earlier literature references the reader is referred to a monograph by *A. Kornberg* (1974) and several recent reviews by *Gefter* (1975), *Geider* (1976), and *Jovin* (1976). Articles on RNA priming of DNA replication have also been published by *Kornberg* (1976) and by *McMacken* et al. (1977).

II. Existence of Several Initiation Mechanisms for DNA Synthesis Involving RNA Primers

Biochemical and genetic evidence in the *Escherichia coli* system suggests that a distinction can be made between two RNA-primed initiation mechanisms: one that initiates DNA synthesis starting at the origin of replication, and another, which seems to be employed during the discontinuous propagation of the DNA chains, and initiates the synthesis of nascent DNA fragments. Still other initiation mechanisms involving RNA primers may be at work in bacteriophage DNA replication. The main types of DNA synthesis for which RNA priming is very probable shall be discussed now.

A. DNA Synthesis Starting from an Origin of Replication

New rounds of replication of the *E. coli* chromosome start at a unique origin of replication, which has been mapped in the vicinity of the *ilv* operon (*Bird* et al., 1972; *McKenna* and *Masters*, 1972; *Hohlfeld* and *Vielmetter*, 1973; *Louarn* et al., 1974). Its position is at about 82 min on the revised genetic map of *E. coli* (*Bachmann* et al., 1976). Initiation of DNA synthesis at this origin is more directly inhibited by rifampicin, a specific inhibitor of DNA-dependent RNA polymerase than by chloramphenicol (*Silverstein* and *Billen*, 1971; *Messer*, 1972). Therefore, the inhibition by rifampicin seems to be due not to the indirect prevention of protein synthesis but to the inhibition of the synthesis of a necessary RNA. One interpretation of these results has been that a functional RNA polymerase transcribes an RNA that may serve as a core for the assembly of the replication proteins (*Lark*, 1972a). Whether this interpretation is true or whether the RNA would actually serve as a primer for extension by DNA polymerase is not yet clear. The latter possibility may be true if it turns out that an RNA that can be isolated covalently linked to high-molecular-weight DNA, is the primer RNA for the initiation of DNA synthesis at the origin. Because of its presumed function, this RNA has been termed oriRNA (*Messer* et al., 1975).

Extrachromosomal elements in *E. coli*, such as plasmids and bacteriophages, to the extent that they depend on the replication machinery of the host, may also make use of this initiation mechanism. The colicinogenic factor Col E1 may be an example. Its replication starts at a unique origin in vivo (*Inselburg*, 1974; *Lovett* et al., 1974) as well as in an in vitro system (*Tomizawa* et al., 1974), and is inhibited by rifampicin, but not by chloramphenicol (*Clewell* et al., 1972; *Tomizawa* et al., 1975). Inhibition of the initiation of replication by rifampicin has been reported for other plasmids as well (see below). By fulfilling the criteria of uniqueness of the origin and inhibition by rifampicin, the replication of bacteriophage λ DNA resembles the DNA synthesis in the host starting from the origin of replication. λ-DNA replication starts at a unique site (*Schnös* and *Inman*, 1970) that has been mapped between genes c II and O (*Stevens* et al., 1971). As shown by genetic analysis, transcription appears to be a necessary prerequisite for the initiation of lambda DNA replication in vivo (*Dove* et al., 1971) and in an in vitro system (*Klein* and *Powling*, 1972). However, it is puzzling that the replication of wild-type λ DNA is inhibited by rifampicin while the replication of a λ-derived plasmid λdv is not (*Hobom* and *Hobom*, 1973). Although this circular dimeric molecule contains two copies of the ori sites, an initiation mechanisms for DNA synthesis different from that used for the wild-type λ DNA seems to be employed for its replication.

B. Discontinuous DNA Synthesis

Once DNA synthesis has started at the origin, further synthesis appears to proceed by a discontinuous mechanism (*Okazaki* et al., 1973). In contrast to the start of new rounds of replication at the origin, discontinuous synthesis is not inhibited by rifampicin (*Lark*, 1972a).

Deoxyribonucleotides are polymerized by all known DNA polymerases in the 5′ to 3′ direction into a new DNA strand. Therefore, in principle, one strand of the double helix, the leading strand, could be synthesized continuously while the other, the lagging strand, would require frequent chain starts. However, it appears as if both strands are synthesized discontinuously (*Okazaki* et al., 1973; *Sternglanz* et al., 1976) and multiple starts of short DNA pieces, called nascent DNA or Okazaki fragments, occur in both strands. Evidence was first obtained in *E. coli* that the Okazaki fragments are primed by RNA (*Sugino* et al., 1972). The experimental approaches to show the possible involvement of RNA will be discussed in detail later, but, whichever RNA polymerase may synthesize these RNA primers, it must be different from the well-known RNA polymerase, since as stated above, discontinuous replication is not affected by rifampicin. In *E. coli* the *dna*G protein is a likely candidate for a primer synthetase. From a study using a temperature-sensitive *dna*G mutant it has been implicated in the synthesis of the primers for the Okazaki fragments (*Lark*, 1972a), and it has been shown to act as a rifampicin-resistant RNA polymerase in vitro, providing the primer RNA for the conversion of G4 single-stranded viral DNA into the double-stranded replicative form (*Bouché* et al., 1975b).

The concept of discontinuous replication requires that three problems be solved: (1) RNA serving as primer has to be removed after it has fulfilled its purpose; (2) the gap left after the removal of the RNA has to be filled in with deoxyribonucleotides; and (3) the ends have to be joined in order to form the continuous strands that constitute natural DNA. In the *E. coli* system there is good evidence that DNA polymerase I is responsible (*Olivera* and *Bonhoeffer*, 1974; *Lehman* and *Uyemura*, 1976) for both the excision of primer RNA and the filling of the gap. The 5′ to 3′ nuclease activity of DNA polymerase I can function to remove the primer RNA, and the polymerase activity can fill the gap. Mutants with defects in the 5′→3′ nuclease function accumulate 10 S pieces to an abnormally high degree (*Konrad* and *Lehman*, 1974), which is a strong indication that this enzymatic activity is indispensable for the conversion of Okazaki pieces into larger DNA.

The final joining of the pieces may be accomplished through the action of polynucleotide ligase. This is strongly indicated by the finding that a thermosensitive *E. coli* mutant defective in ligase activity accumulates Okazaki pieces upon a shift to the nonpermissive temperature (*Konrad* et al., 1974).

Recently, however, a different mechanism has been discovered that can give rise to the transient formation of short fragments (*Tye* et al., 1977). DNA polymerases can incorporate in vitro dUTP, the natural precursor of dTTP, into DNA (*Bessman* et al., 1958). However, uracil is not normally found in DNA. This is because in normal cells the level of dUTP is kept very low

by the enzyme dUTPase, and, in addition, cells have excision–repair systems
that detect and remove foreign bases from DNA. It was observed in mutants
of *E. coli*, originally called *sof*, that radioactive pulse-labeling of the cellular
DNA yielded nascent fragments 5–10 times smaller than normal Okazaki fragments (*Konrad* and *Lehman*, 1975). *sof* mutants have been found to be identical
to *dut* mutants, which are deficient in dUTPase (*Tye* et al., 1977). Therefore,
these mutant cells contain a higher than normal ratio of dUTP to dTTP, and
more dUTP than usual appears to be incorporated into their DNA. The uracil
residues are quickly removed, possibly by an excision–repair mechanism like
that proposed by *Lindahl* (1976). However, during the repair there are temporary
breaks introduced into the DNA that may yield small fragments in pulse-labeling
experiments. In view of these results it cannot be excluded that, also in dut^+
cells, a significant proportion of the Okazaki fragments comes from excision–
repair and not necessarily from de novo priming events, and that perhaps
all of the Okazaki fragments from the leading strand and at least some of
the fragments from the lagging strand may arise from the repair reaction.

In eukaryotic cells all DNA, including viral DNA, appears to be synthesized
discontinuously (for reviews see *Taylor*, 1974; *Edenberg* and *Huberman*, 1975;
Winnacker, 1975). Nascent fragments isolated from nuclei of polyoma-infected
cells are smaller than the bacterial Okazaki pieces, about 4 S in size, and they
are primed by an RNA for which the name initiator RNA (iRNA) has been
proposed (*Reichard* et al., 1974). However, it has not yet been possible to find
out which of the several RNA polymerases present in eukaryotic cells functions
in the primer synthesis.

C. Replication of Bacteriophage DNA

1. Bacteriophages Containing Double-Stranded DNA

In contrast to the difficulties encountered in studying the replication of such
a large DNA as the *E. coli* chromosome, bacteriophages offer obvious advan-
tages as model systems. By studying the replication of phages like T4, T7
or λ one could hope to gain knowledge about replication mechanisms that
may provide clues also applicable to the replication of the host chromosome.
Replication starts in vivo at a unique origin on the chromosomes of the bacteriophages T4 (*Mosig*, 1970), T7 (*Dressler* et al., 1972), λ (*Schnös* and *Inman*, 1970),
and P2 (*Schnös* and *Inman*, 1971). Short DNA intermediates, taken as an indication for the discontinuous mode of chain propagation, have been reported
for T4 (*Sugino* and *Okazaki*, 1972), for T7 (*Miller*, 1972), and for P2 (*Kurosawa*
and *Okazaki*, 1975). An indication for the RNA priming of T4 replication,
obtained by density shift experiments (*Buckley* et al., 1972; *Speyer* et al., 1972),
should be regarded with reservation (see below). However, RNA primers have
also been suggested for the discontinuous replication of phages T4 and P2
on the basis of evidence obtained with the polynucleotide kinase and the spleen
exonuclease methods (*Okazaki* et al., 1975b).

In vitro replication systems reconstituted from purified proteins for T4 replication (*Morris* et al., 1975; *Alberts* et al., 1977) and for T7 replication (*Masker*

and *Richardson*, 1976; *Scherzinger* and *Klotz*, 1975) did not yield conclusive evidence for RNA priming. However, a requirement for all four ribonucleoside triphosphates for optimal T4 replication as well as for T7 replication may indicate that RNA synthesis precedes DNA synthesis.

In addition to rifampicin inhibition mentioned above, evidence for RNA priming of λ DNA replication has been obtained through the functional analysis of short λ DNA fragments cloned into Col E1 plasmid vectors. That analysis revealed that λ replication requires RNA transcription that runs into the segment carrying the t_o terminator (= oop-RNA terminator). λ-specific, gene O and P-dependent, replication (in temperature-sensitive pol A$^-$ cells at the nonpermissive temperature) does not occur when this transcription is blocked either by deletion of the promoter part or by insertion of an extra terminator piece (carrying the λt_{R1} signal) between the promoter and the t_o terminator. Inversion of the t_o terminator fragment with respect to the promoter fragment will also abolish the λ-specific replication. However, when a DNA fragment containing both the promoter and the t_o terminator is inserted inversely with respect to the rest of the plasmid, λ-specific replication does occur (*Lusty, M.* and *Hobom, G.*, 1977 personal communication).

2. Bacteriophages Containing Single-Stranded DNA

In prokaryotic systems the most convincing evidence for RNA priming of DNA synthesis has come from work on the replication of single-stranded DNA containing bacteriophages such as the filamentous phages M13 and its relatives fd and f1, and the spherical phages φX174 and the related phages G4 and S13. The biology of the filamentous phages has been reviewed by *Marvin* and *Hohn* (1969), and their replication by *Ray* (1977). The biology of the spherical phages has been reviewed by *Sinsheimer* (1968), and their reproduction recently by *Denhardt* (1977). For both phage classes the reproductive cycle comprises three stages: At first the circular single-stranded viral strand is injected into the host and immediately used as a template for the synthesis of a complementary strand (SS→RF II conversion). The product, named parental replicative form II (RF II), contains the circular, intact viral strand and, annealed to it, an almost complete complementary strand. Yet, a small gap between the 5′ and the 3′ end remains in the complementary strand. In vitro, this gap can be filled in by DNA polymerase I and the adjacent ends can be joined by ligase. The same mechanism probably works in vivo, too. The resulting double-stranded circular molecule, having two covalently closed DNA strands, is called replicative form I (RF I). During the second stage the duplex RF is multiplied (RF→RF). The final stage is the production of progeny single strands, for which a rolling circle type mechanism has been proposed. The duplex RF, which has been nicked in its viral strand by a phage-coded specific endonuclease, serves as primer template for the synthesis of viral strands (RF→SS).

Although the overall scheme of reproduction is similar for both the filamentous and the spherical phages, both classes differ in the replication mechanisms of the single stages. Thus far, the SS to RF conversion of M13 viral DNA is the most clear-cut example for RNA priming of DNA synthesis. The reaction is inhibited by rifampicin in vivo (*Brutlag* et al., 1971) and in vitro (*Wickner, W.* et al., 1972). No inhibition is observed in a rifampicin-resistant RNA polymerase mutant. All four ribonucleoside trophosphates are required for optimal in vitro conversion and are supposedly incorporated into the primer RNA (*Wickner, W.* et al., 1972). In contrast, the SS→RF conversion of φX174 and that of the related phage G4, are not inhibited by rifampicin (*Schekman* et al., 1972, 1974). Yet, RNA priming of the reaction is suggested by several lines of evidence. First, label transfer from [α-^{32}P] deoxynucleotides to ribonucleotides has been reported for φX174 (*Schekman* et al, 1972), and a piece of RNA covalently linked to the 3' end of the complementary strand of G4-RF II synthesized in vitro has been isolated. The *dnaG* protein, characterized as a rifampicin-resistant RNA polymerase, is responsible for the synthesis of this primer RNA in vitro on DNA unwinding protein-covered G4 viral strands (*Bouché* et al., 1975 b). A comparison of the in vitro replication of phages M13, φX174, and G4 and of their roles, as probes for the investigation of the replication machinery of E. coli has been published by *A. Kornberg* (1977).

D. DNA Replication in Small Animal Viruses

The replication of the double-stranded DNA of small animal viruses such as polyoma and SV40 initiates at fixed origins of replication (*Crawford* et al., 1974; *Nathans* and *Danna*, 1972) by a yet unknown mechanism. The chain propagation of both strands follows a discontinuous mode of DNA synthesis (*Fareed* and *Salzman*, 1972; *Pigiet* et al., 1973). Studies of the discontinuous replication of polyoma DNA in vitro indicated strongly that the majority of the 4 S DNA chains are initiated by RNA (*Magnusson* et al., 1973; *Hunter* and *Francke*, 1974). Nascent fragments during SV40 replication are initiated by RNA as well (*Kaufmann* et al., 1977).

E. DNA Synthesis by Reverse Transcriptase

Synthesis of the proviral DNA of a number of RNA tumor viruses by reverse transcriptase may be primed by transfer RNA molecules tightly associated with the RNA genomes inside the virion. It was first shown for Rous sarcoma virus that a 4 S RNA that remains complexed with the viral RNA after disruption of the virion by detergent serves as a primer for DNA synthesis by reverse transcriptase in vitro (*Dahlberg* et al., 1974). This 4 S RNA was identified as cellular tryptophanyl-tRNA (*Harada* et al., 1975). Although tRNA^{Trp} seems to be a unique genome-associated primer for the avian myeloblastosis virus (AMV) other tRNAs have been found to serve as primers for other viruses. For example, one particular isoaccepting species of tRNA^{Pro} seems to function as a primer for Moloney murine leukemia virus (*Peters* et al., 1977) and tRNA^{Pro} has also been found to be the primer 4 S RNA for AKR, Friend, and Rauscher murine leukemia viruses, for simian sarcoma virus, and for feline leukemia virus (*Waters,* 1975; *Waters* and *Mullin*, 1976). The role of tRNA species as primers for the transcription into DNA of RNA tumor virus genomes in vitro has been reviewed recently by *Taylor* (1977), *Dahlberg* (1977), and *Waters* and *Mullin* (1977).

III. Evidence for RNA Priming of DNA Synthesis

A. Direct Evidence for RNA Priming

To obtain direct evidence that RNA is serving as the primer for DNA synthesis one approach often used is to demonstrate a covalent linkage between the RNA and the DNA. However, mere demonstration of covalent linkage may not be sufficient to make a primer function plausible. It must be shown, in addition, that the RNA is linked to the 5′ end of the DNA. The following methods have been employed for demonstrating covalent RNA-DNA linkage:

1. Comigration under Denaturing Conditions of RNA and DNA Labeled with Different Isotopes

The RNA that primes in vitro synthesis of the complementary strand of bacterio-phage G4-DNA was shown to cosediment with the newly synthesized DNA in sucrose gradients made up in pure formamide (*Bouché* et al., 1975b). Forma-mide at high concentrations disrupts hydrogen bonds that may hold together RNA-DNA hybrids and DNA-DNA hybrids (*McConaughy* et al., 1969; *Casey* and *Davidson*, 1977). Therefore, the cosedimentation of the primer RNA labeled with ^{32}P together with the newly synthesized DNA, which was labeled with ^{3}H, indicated covalent linkage between RNA and DNA. A single cleavage site in the synthesized G4 replicative form II DNA for the restriction endonu-clease Eco RI served as a reference point to show that the RNA fragment was covalently bound to the DNA at the 5′ end. Similarly, it was shown that, under denaturing conditions, RNA which was synthesized by DNA-dependent RNA polymerase of *E. coli* on φX174 viral strands served as the primer for covalent extension by DNA polymerase II from human KB cells, and sedimented together with the newly synthesized DNA in a sucrose gradient containing a mixture of deuterium oxide and dimethylsulfoxide, which supposedly disrupts hydrogen-bonded RNA-DNA hybrids, as well (*Keller*, 1972).

2. Density Shift in Isopycnic Centrifugation Experiments

Evidence for covalent linkage between RNA and DNA in nascent fragments isolated from *E. coli* has been derived from the observation that in cesium sulfate isopycnic gradients the buoyant density of nascent DNA pieces labeled by a very brief pulse with [^{3}H] thymidine is greater than that of the bulk of the DNA. Furthermore, some [^{3}H]uridine-labeled RNA bands in the DNA region in such gradients (*Sugino* et al., 1972; *Okazaki* et al., 1973; *Hirose* et al., 1973). Similarly, in Ehrlich ascites tumor cells (*Sato* et al., 1972), in human lymphocytes (*Fox* et al., 1973), in chinese hamster ovary cells (*Taylor, J.H.* et al., 1975), the slime mold *Physarum polycephalum* (*Waqar* and *Huberman*,

1973), in HeLa cells (*Blinkerd* and *Toliver*, 1974; *Olgiati* et al., 1976), and in polyoma (*Magnusson* et al., 1973; *Hunter* and *Francke,* 1974; *Sadoff* and *Cheevers,* 1973), an increased density of nascent fragments was observed. However, others were unable to detect a higher density of nascent DNA isolated from mouse myeloma cells (*Berger* and *Huang,* 1974) or mouse P-815 cells (*Gautschi* and *Clarkson,* 1975), and upon reexamination of the nascent DNA isolated from *E. coli* in the laboratory, where the density shifts originally were observed and interpreted as covalent RNA-DNA linkages, these shifts could not be reproduced (*Ogawa* et al., 1977). Therefore, evidence for covalent linkage between RNA and DNA based on density shift experiments should be considered weak, unless the possibility of the reassociation of RNA and DNA during centrifugation is rigorously eliminated, for instance, by denaturation, and centrifugation of the nascent fragments in the presence of a sufficiently high concentration of formaldehyde to prevent reassociation (*McGhee* and *von Hippel,* 1977). It has been shown that without such treatment reannealing can occur very readily and simulate a covalent attachment of RNA to DNA (*Mendelsohn* et al., 1975; *Pearson* et al., 1976; *Probst* et al., 1974; *Reichard* et al., 1974).

3. Label Transfer Experiments

Covalent linkage between RNA and DNA has been inferred from the phosphate transfer from deoxyribonucleotides to ribonucleotides at the RNA-DNA junction. For these experiments usually a variation of the nearest neighbor analysis technique is employed, which was originally described for DNA by *Josse* et al. (1961). At the junction between RNA and DNA a radioactively labeled phosphorus is introduced by incorporation of an $[\alpha-^{32}P]$ dNTP as the first deoxyribonucleotide of the DNA chain. This radioactive phosphate group is transferred to the last ribonucleotide of the primer RNA either in the 2′ or 3′ position upon alkaline hydrolysis and allows the unambiguous identification of the donor deoxynucleotide and the acceptor ribonucleotide. Using this approach ribo-deoxyribo linkage has been shown in nascent fragments isolated from toluenized E. coli cells (*Sugino* and *Okazaki,* 1973), in lysates of various mammalian cell lines (*Waqar* and *Huberman,* 1975; *Waqar* et al., 1975), and in nascent DNA isolated from nuclei of the slime mold *Physarum polycephalum* injected with radioactive precursors (*Waqar* and *Huberman,* 1975a). RNA-DNA linkages have also been demonstrated by label transfer experiments in small fragments synthesized in vitro on polyoma virus DNA, (*Magnusson* et al., 1973) and on SV40 DNA (*Anderson* et al., 1977).

For the nascent fragments of permeabilized *E. coli* a predominant ^{32}P transfer from dCTP to UMP or from dGTP to UMP was originally reported (*Sugino* and *Okazaki,* 1973). Reexamination of these experiments, though, could not confirm these results (*Okazaki* et al., 1975b). Instead, in *E. coli* nascent fragments also, all sixteen possible combinations seem to occur (*Ogawa* et al., 1977) with roughly equal frequency, which would agree with results obtained in nascent fragments isolated from *E. coli* lysates (*Ramareddy* et al., 1975), from the slime mold *Physarum polycephalum* (*Waqar* and *Huberman*, 1975a), and human lymphocytes (*Tseng* and *Goulian*, 1975). For polyoma a slight preference of transfer from dC has been reported (*Magnusson* et al., 1973; *Hunter* and *Francke*, 1974). However, repetition of these experiments with an improved technique did not confirm that transfer from one deoxynucleotide is more frequent than from the three others in polyoma nascent fragments, either by label transfer experiments or by polynucleotide kinase labeling of the 3′-terminal deoxynucleotide after removal of the primer RNA by alkali (*Pigiet* et al., 1974). Specific transfer of label from [α-^{32}P] dNMP to 2′ (3′) rAMP was observed at the primer RNA-DNA junction present in the in vitro-synthesized complementary strand of bacteriophage M13 DNA (*Schekman* et al., 1972). Also, transfer occurred more frequently to rG and rA than to rU and rC in the RNA-DNA junction of the in vitro-synthesized complementary strand of φX174 bacteriophage DNA (*Schekman* et al., 1972). Label transfer may appear a convincing way of demonstrating an RNA-DNA junction. However, the use of deoxyribonucleoside triphosphates of very high specific radioactivity, which may be necessary to detect the relatively rare label transfer, for instance, once every 2000 bases in nascent fragments of *E. coli* or only once every 6000 bases during the synthesis of the M13 RF II, includes the risk of picking up artifacts arising, e.g., from misincorporation of ribonucleotides into DNA. Furthermore, as pointed out by *Hartman* and *Werner* (1977), some of the label added as [α-^{32}P] deoxyadenosine triphosphate could end up in RNA precursors because of nucleotide turnover during the incubation period, and RNA synthesized from these precursors may yield 2′,3′ ribonucleotides after alkaline hydrolysis indistinguishable from those that arise from direct label transfer.

4. Evidence Derived from Phosphorylation of Terminal Nucleotides

An elegant method for demonstrating linkage of RNA to the 5′ end of DNA has been devised using polynucleotide kinase (*Hirose* et al., 1973). 5′-OH ends of nascent DNA are blocked by primer RNA and are therefore inaccessible to enzymatic phosphorylation by T4 polynucleotide kinase. They become accessible, though, after the ribonucleotides have been removed by alkali treatment and can then be phosphorylated with radioactive phosphorus from [γ-^{32}P]ATP. By this method, the ends can be tagged and counted. The method is very sensitive but has the disadvantage that the reaction conditions have to be chosen carefully in order to minimize the exchange of phosphate between ATP and 5′-phosphoryl-terminated DNA fragments, a reaction that is also catalyzed by the kinase (*van de Sande* et al., 1973). The exchange reaction can be largely suppressed if the incubation temperature during the kinase reaction is kept at 0° C (*Okazaki* et al., 1975a); however, even under these conditions, as much as 3% of any 5′-phosphoryl-terminated DNA may exchange its terminal phosphate (*Ogawa* et al., 1977). Therefore, if the amount of free DNA pieces is large relative to the amount of nascent, RNA-containing pieces, the exchange reaction can severely impair accurate measurement of the RNA-DNA linkages by the kinase method.

5. The Spleen Exonuclease Assay

An alternative method for assaying RNA-linked nascent DNA fragments has been developed using spleen exonuclease (*Okazaki* et al., 1975b; *Kurosawa* et al., 1975). Only DNA having a free 5'-OH terminus is degraded by spleen exonuclease. Nascent fragments containing an RNA primer would expose a free 5'-OH deoxyribonucleotide end after removal of the ribonucleotides. The presence of RNA covalently linked to DNA at its 5' end therefore would be indicated by the liberation of deoxyribonucleotides after alkali treatment of the nascent fragments. Interference from contaminating DNA fragments having free 5'-OH ends can be prevented by phosphorylation of these fragments with polynucleotide-kinase before the alkali treatment. The spleen exonuclease assay was used successfully to demonstrate the presence of RNA at the 5' end of nascent fragments isolated from *E. coli* cells (*Okazaki* et al., 1975b).

6. Inclusion of Primer RNA in a DNA Strand by Joint Action of T4-DNA Polymerase and T4 Ligase

Evidence that RNA serves as primer for the in vitro synthesis of the complementary DNA strand of the replicative form II (RF II) of bacteriophages M13 and φX174 has been obtained by an ingenious approach: When T4-DNA polymerase and T4 ligase were used to convert in vitro the replicative form II DNA of these phages (which has a gap in the complementary strand) into the covalently closed, duplex replicative form I (RF I), this RF I was found to be alkali sensitive (*Westergaard* et al., 1973). RNA that served as primer was evidently preserved and incorporated into the complementary strand because the T4-DNA polymerase lacks a $5' \rightarrow 3'$ exonuclease function that appears to be needed for the excision of the primer RNA during the gap-filling process. The RNA is sealed covalently into the complementary strand by T4 ligase, which is able to join DNA and RNA segments (*Lehman*, 1974). In contrast, no covalently closed RF I was formed when *E. coli* ligase, which cannot join DNA and RNA, was used instead of T4 ligase. On the other hand, the RF I formed by the joint action of *E. coli* DNA polymerase I and *E. coli* ligase was alkali insensitive. *E. coli* DNA polymerase I with its $5' \rightarrow 3'$ exonucleolytic activity apparently removed the primer RNA while filling the gap, and thereby created the conditions necessary for the sealing of the strand by *E. coli* ligase, which can only join adjacent deoxynucleotide ends. Although this approach has not yet found widespread application in other systems, it has proved its potential with the demonstration of the RNA-DNA link in the in vitro-synthesized RF IIs of bacteriophages M13 and φX174 (*Westergaard* et al., 1973). The polymerase ligase method has recently been used to probe for ribonucleotides in M13 RF II isolated from *E. coli* RS 5052, a mutant strain deficient in the $5' \rightarrow 3'$ exonuclease function of DNA polymerase I. From the analysis it appears as if RF II made under these conditions contains ribonucleotides in both strands (*Dasgupta*, 1977).

B. Indirect Evidence for RNA Priming

1. Inhibition of DNA Synthesis by Drugs Known to Inhibit Transcription

a) Rifampicin

Rifampicin, a derivative of the natural antibiotic rifamycin SV, inhibits the growth of *E. coli* by blocking the activity of DNA-dependent RNA polymerase (*Hartmann* et al., 1967; *Wehrli* and *Staehelin,* 1971; *Riva* and *Silvestri,* 1972). The enzyme appears to be the only target of the drug (*Tocchini-Valentini* et al., 1968; *Ezekiel* and *Hutchins,* 1968). So far no other target for rifampicin has become known. Therefore, if it can be shown that DNA synthesis is affected by rifampicin directly and not indirectly via its effect on translation i.e. that the synthesis of a mRNA coding for a protein, which is required for DNA synthesis, is prevented, the inference may be justified that RNA polymerase or at least its β subunit, with which the drug interacts (*Heil* and *Zillig,* 1970), plays a role in the DNA synthesis being inhibited, presumably for the synthesis of a primer RNA. Usually the inhibition by rifampicin is contrasted to that of chloramphenicol to distinguish whether transcription is the crucial step or protein synthesis. Thus, it was shown that rifampicin, added at a time in the replication cycle when chloramphenicol was no longer inhibitory, would prevent the initiation of new rounds of replication from the origin in *E. coli* cells synchronized by amino acid starvation (*Lark,* 1972 b) or by the filtration technique (*Messer,* 1972). For a number of plasmids and phages, inhibition of replication by rifampicin has been suggestive evidence for the requirement of RNA synthesis. Thus, replication of F-episomes (*Bazzicalupo* and *Tocchini-Valentini,* 1972; *Kline,* 1972, 1973; *Hiraga* and *Saitoh,* 1975), the replication of Col E1 in vivo in the presence of chloramphenicol (*Clewell* et al., 1972) and in vitro in an *E. coli* extract (*Sakakibara* and *Tomizawa,* 1974a) or in plasmolyzed cells (*Staudenbauer,* 1975), and the reproduction of the small minicircular DNA in *E. coli* 15 (*Messing* et al., 1972) are inhibited by rifampicin. Conversion of the single-stranded DNA of bacteriophage M13 into double-stranded RF II is strongly inhibited by rifampicin in vivo in an *E. coli* wild-type strain but not in a rifampicin-resistant RNA polymerase mutant strain (*Brutlag* et al., 1971). The conversion is also inhibited in vitro (*Wickner, W.* et al., 1972). However, a wild-type extract inactivated by the addition of rifampicin will resume synthesis of the complementary strand when purified rifampicin-resistant RNA polymerase is added (*Zechel,* unpublished results).

A rifampicin-sensitive step seems to precede the synthesis of the complementary strand during the second stage of the M13 reproductive cycle, the RF to RF replication (*Fidanián* and *Ray,* 1974). Perhaps RNA polymerase provides a primer RNA during this step, too. The last stage of M13 reproduction, i.e. the synthesis of the viral strands, is also inhibited by rifampicin faster than by chloramphenicol, and on the basis of this differential inhibition, an involvement of RNA polymerase in the viral strand synthesis was suggested (*Staudenbauer* and *Hofschneider,* 1972). However, rifampicin appears to affect the viral-strand synthesis indirectly by inhibiting the expression of the gene

5 protein whose concentration inside the cell is crucial for viral-strand production (*Fidanián* and *Ray*, 1974).

Inhibition of phage λ DNA replication by rifampicin and, consequently, the involvement of RNA polymerase in its initiation has been mentioned above.

b) Streptolydigin

Streptolydigin, like rifampicin, binds to the β-subunit of RNA polymerase (*Heil* and *Zillig*, 1970). The binding is reversible and affects a different site than rifampicin (*Ghysen* and *Pironio*, 1972; *Iwakura* et al., 1973). While rifampicin inhibits initiation of transcription (*diMauro* et al., 1969), streptolydigin inhibits subsequent chain elongation of the RNA (*Schleif*, 1969; *Cassani* et al., 1971). Streptolydigin has been used as an alternative for rifampicin for probing that RNA polymerase plays a role in the initiation for new rounds of replication at the origin of the *E. coli* chromosome (*Lark*, 1972b) and in the conversion of M13 viral DNA to RF II in vitro (*Schekman*, et al., 1972). The replication of Col E1 DNA studied in the presence of chloramphenicol is also inhibited by streptolydigin but it was found to be at least 100 times less sensitive than cellular RNA synthesis. In contrast, the inhibition of DNA synthesis by rifampicin was approximately the same as the inhibition of cellular RNA synthesis (*Clewell* and *Evenchik*, 1973). The authors speculate that RNA polymerase may be modified for primer synthesis, which makes it less susceptible to inhibition by streptolydigin but not by rifampicin. However, the Col E1 DNA replication in cell-free extracts is fairly sensitive to streptolydigin while relatively high concentrations of the drug are necessary to inhibit the M13 SS to RF conversion. Therefore, the sensitivity to the drug may also be related to the properties of the template.

c) Actinomycin D

This drug is widely used as an inhibitor of transcription. However, it inhibits not only transcription but, although less effective, also replication because it acts by intercalating into duplex DNA and binding to deoxyguanosine (*Sobell* and *Jain*, 1972). Inhibition of the conversion of M13 or φX174 viral strands in vitro by concentrations of actinomycin D, just high enough to inhibit transcription but still low enough not to affect replication substantially, was taken together with other evidence to be discussed below to argue that both reactions contain a transcriptional step, presumably the synthesis of an RNA primer (*Schekman* et al., 1972). The inhibitory effect of actinomycin D on the replication of Col E1 DNA in the presence of chloramphenicol, which was lower than that of rifampicin but comparable to that of streptolydigin, was interpreted as additional evidence for the RNA priming of this reaction (*Clewell* and *Evenchik*, 1973). Although actinomycin D is the only drug among those discussed that inhibits transcription by bacterial as well as animal polymerases, the possibility that it already interferes with replication at low concentrations makes it appear less suited for probing the role of transcription for priming of DNA synthesis.

2. Other Indirect Evidence

a) Requirement for a Functional RNA Polymerase in Bacterial Systems

The requirement for a functional RNA polymerase for the synthesis of M13 parental RF in vitro has been demonstrated not only through the inhibition by drugs but also by studies of an RNA polymerase mutant and anti-RNA polymerase antibodies. An extract prepared from a temperature-sensitive RNA polymerase mutant converted M13 viral strands to RF II at the permissive temperature but failed to do so at the nonpermissive temperature (*Geider* and *Kornberg,* 1974). From the fact that the conversion in vitro of fd, (or M13 DNA) but not that of φX174 DNA, could be inhibited by antibodies raised against purified RNA polymerase, independent evidence was obtained that RNA polymerase is required for the conversion of fd single-strand, but is dispensible for φX174 viral strand conversion (*Wickner, R.B.* et al., 1972).

3. Requirement of Four Ribonucleoside Triphosphates for Replication

A requirement of ribonucleoside triphosphates (rNTP) for replication has been observed in many systems and has been taken as an indication that RNA synthesis is coupled to DNA synthesis and may serve the purpose of providing the primer. Since cells are normally impermeable for the highly charged nucleoside triphosphates, these studies are restricted to in vitro systems either in cell extracts or in cells permeabilized for rNTPs by various treatments. In such crude systems it is usually only possible to observe a more or less pronounced stimulation of replication by rNTPs because, despite dialysis or gel filtration steps, there may be still enough rNTP present to meet the possibly extreme low-level requirement of primer RNA synthesis. An up to 20-fold stimulation of the replicative DNA synthesis by ATP (e.g. *Moses* and *Richardson,* 1970) is common to the nucleotide-permeable system used to study replication of *E. coli* phages λ or T4 (several systems are described in *R.B. Wickner,* 1974). Other nucleoside triphosphates can partially substitute for ATP (*Vosberg* and *Hoffmann-Berling,* 1971, *Pisetsky* et al., 1972). However, no evidence has been presented that the ATP requirement is connected to the synthesis of a primer. ATP may be needed just to provide energy for one or several energy-requiring reactions. For instance, ATP is required for the formation of an initiation complex between DNA polymerase III, copolymerase III, and the primer terminus (*Wickner, W.* and *Kornberg,* 1973). Several of the proteins involved in the replication process have been shown to possess ATPase activity, notably the *dna*B protein (*Wickner, S.* et al., 1974), a replication factor from *E. coli* for φX174 viral strand conversion called Y (*Wickner, S.* and *Hurwitz,* 1975c), the rep-protein, which exhibits a single-stranded DNA-dependent ATPase activity during its presumed function of melting duplex DNA in the replicative fork (*Scott* et al., 1977), and DNA gyrase, which introduces superhelical turns into circular DNA molecules (*Gellert* et al., 1976). Two other enzymes, termed DNA helicases, have recently been described, which bind to single-stranded DNA regions and unwind adjacent double-stranded regions. This process is dependent on the presence of ATP which is hydrolyzed during the unwinding step. DNA helicase I may not be involved in DNA replication. DNA helicase II is probably identical with the rep-protein (*Abdel-Monem* et al., 1976; 1977a, b).

A clear requirement for all four ribonucleoside triphosphates has been shown for the replication of Col E1 in plasmolyzed *E. coli* cells (*Staudenbauer*, 1975) and in a crude cell-free extract (*Sakakibara* and *Tomizawa*, 1974a). Since, the rifampicin sensitivity of Col E1 replication discussed above, the presence of oligoribonucleotide stretches in Col E1 DNA isolated from cultures grown for several generations in the presence of chloramphenicol (*Blair* et al., 1972) and the association of RNA with a 6 S replication intermediate isolated in vitro (*Sakakibara* and *Tomizawa*, 1974b) indicate that RNA primers may initiate Col E1 replication, the requirement for ribonucleotides does fit the picture.

Up to 20-fold stimulation by ribonucleoside triphosphates has been observed for the replication of bacteriophage T7-DNA in vitro in extracts from T7 infected *E. coli* cells (*Strätling* et al., 1973; *Hinkle* and *Richardson*, 1974). T7-DNA replication by purified enzymes is also stimulated but not dependent on ribonucleoside triphosphates (*Hinkle* and *Richardson*, 1975; *Scherzinger* and *Klotz*, 1975). However, recent studies suggest that the ribonucleoside triphosphates play a role as energy sources for the energy-requiring unwinding of the duplex DNA rather than being precursors for primer RNA synthesis (*Kolodner* and *Richardson*, 1977). The replication of T4-DNA in a cellophane-disc system prepared from T4-infected *E. coli* shows a threefold stimulation when ATP is added (*Imae* and *Okazaki*, 1976). With the bacteriophage T4-DNA replication apparatus reconstructed from the purified protein products of the phage T4 genes 41, 43, 44, 45, 62, 32, and the X protein, extensive DNA synthesis on duplex and on single-stranded templates can be achieved. However, initiation of DNA chains de novo on a single-stranded template is found only if ribonucleoside triphosphates are present (*Alberts* et al., 1975; 1977; *Morris* et al., 1975). In view of the fact that rifampicin-resistant T4 replication proceeds by a discontinuous mechanism (*Okazaki* and *Okazaki*, 1969), the ribonucleotides may be used for the synthesis of RNA primers for the Okazaki pieces.

An absolute requirement for all four ribonucleoside triphosphates has been found for the in vitro conversion of the viral SS to RF of the filamentous phages M13 and fd (*Wickner, W.* et al., 1972; *Wickner, R.B.* et al., 1972), a result very much in agreement with the notion that RNA polymerase synthesizes an RNA primer for this reaction.

A stimulation of G4-RF II synthesis in vitro by ATP and GTP has been observed (*Zechel* et al., 1975), and incorporation of all four ribonucleoside triphosphates into an isolatable primer RNA has been described (*Bouché* et al., 1975a). Nevertheless, the requirement of ribonucleoside triphosphates for the in vitro SS to RF conversion of the viral strand of φX174 and its relatives is still controversial. There seems to be agreement over the principal need of ATP for optimal conversion. However, ATP appears to be involved in one or more reactions preceding the primer synthesis. Thus, it has been described that the formation of a replication intermediate from the φX174 viral template and five proteins, which precedes the *dna*G RNA polymerase action, requires ATP (*Wickner, S.* and *Hurwitz*, 1974; *Weiner* et al., 1976). It has not been established, in which step(s) the ATP is necessary, but one ATP-dependent reaction during the formation of this replication intermediate could be the association of the *dna*B and the *dna*C (D) proteins. Purified *dna*B protein has been shown to associate with pure *dna*C protein only in the presence of ATP (*Wickner, S. and Hurwitz,* 1975a, b).

A strong stimulation by GTP, UTP, and CTP of φX174 RF II synthesis in vitro has been demonstrated by one group (*Schekman* et al., 1974) and the transfer of label from [α-^{32}P] deoxyribonucleotides to all four ribonucleotides appears to support the RNA-primer hypothesis (*Schekman* et al., 1972). However, others have found that no other ribonucleoside triphosphates were necessary besides ATP for the conversion of φX174 viral strands to the replicative form II in vitro (*Wickner, R.B.* et al., 1972; *Wickner, S.* and *Hurwitz,* 1974). In light of a recent report that the *dna*G protein can synthesize in vitro an oligonucleotide primer on the DNAs from phages G4 or ST-1, two relatives of φX174, using either deoxyribonucleoside triphosphates or ribonucleoside triphosphates (*Wickner, S.,* 1977), it is conceivable that either substrate may be sufficient for the synthesis of a suitable primer for the conversion of φX174 viral strands to RF II in vitro, and that ribonucleoside triphosphates other than ATP may not be necessary for the synthesis of a primer when deoxyribonucleoside triphosphates are present. The stimulation of primer formation by ADP is an observation that appears to deserve some further attention.

IV. The Structure of Primer RNAs

While direct and indirect evidence suggesting RNA as primer for DNA synthesis is convincing for several systems, notably the M13 and G4 SS to RF conversion in vitro, the polyoma nascent DNA synthesis in vitro, and the discontinuous synthesis in various eukaryotic systems, the isolation and characterization of the primer RNAs has turned out to be extremely difficult. One could conceive of a number of reasons why primer RNAs are so elusive. For example, their function as primer may require them to be very unstable. They are presumably one-way molecules ear-marked for rapid degradation. RNA primers in the nascent pieces, the intermediates during discontinuous replication, ought to be removed as quickly as possible after they have served their purpose. Otherwise they would become obstacles to the formation of a continuous DNA strand. Furthermore, primer RNAs may be very small RNA species, making it difficult to detect and identify them within the cellular RNA pool. Also, the average amount of some RNA primers, e.g. at the origin may be small because they are present only during certain stages of the reproductive cycle. This means that very sensitive methods are required for their detection. Nevertheless, with the refinement of the methods it has become possible to isolate a number of RNA species that are likely to be primer RNAs, and to obtain data with respect to their size and, in some cases, even to their sequences. In early attempts to estimate the size of the primer RNA on Okazaki pieces from *E. coli,* values of 50 to 100 nucleotides were reported. These estimates were based mainly on gel filtration studies combined with density centrifugation analyses of the RNA-DNA copolymers isolated from *E. coli* after short pulses with radioactively labeled thymidine and/or uridine (*Okazaki* et al., 1973). However, others were unable to confirm the presence of RNA in nascent fragments isolated from the same *E. coli* strain, or even from *E. coli* mutants that accumulate nascent pieces (*Eichler, Thorner* and *Lehman* cited in *Uyemura* et al., 1976). These mutants were thought to provide better chances for the detection of the primer RNA because they are defective in the 5′→3′ exonuclease function of DNA polymerase I, which is presumed to be responsible for the removal of the RNA from nascent pieces. Upon reexamination of the problem in the original laboratory with improved techniques, and this time including in the studies the conditionally lethal DNA polymerase mutants, the authors again concluded that RNA is present in the nascent pieces. However, in contrast to the earlier findings, the size of the primer oligoribonucleotides was estimated to range from mono- to trinucleotides (*Ogawa* et al., 1977).

Another partly characterized RNA that may be a primer is the ori-RNA, which is believed to initiate the DNA at the origin of replication. It has been estimated to be not more than 500 nucleotides long. This estimate is based on the analysis in polyacrylamide gels of the RNA part that remained after the DNA fraction of the oriRNA-DNA copolymers, isolated from [³H]uridine-pulse-labeled *E. coli* cells, had been digested with DNAse (*Messer* et al., 1975).

Supercoiled circular Col E1 DNA containing a single stretch of RNA in either strand can be isolated after extensive replication in the presence of chloramphenicol, which prevents the replication of the host's DNA but does not interfere with Col E1 DNA replication (*Blair* et al., 1972). The ribonucleotide stretches occur with equal probability in either the light or the heavy strand of the Col E1 DNA (*Williams* et al., 1973). Analysis of the RNA showed that the light-strand RNA segment consists of 38 ribonucleotides. The base composition is 17 G, 5 A, 8 C, and 8 U, and the sequence of this RNA segment has been partially determined: 5′...p(rG)p(rG)p(rG)p(rG)p(rA)p(rC)...3′. The deoxyribonucleotide-ribonucleotide junction at the 5′ end of this RNA segment is specifically ...p(dC)p(rG)... The RNA segment associated with the heavy strand consists of 15 ribonucleotides with a base composition of 5 G, 2 A, 4 C, and 4 U. The partial sequence of its 5′ terminus reads 5′...p(rG)p(rA)p(rA)-p(rU)p(rG)...3′ and its deoxyribonucleotide-ribonucleotide junction is uniquely...p(dA)p(rG)... (*Helinski* et al., 1975; *Helinski*, 1976). The RNA segments appear to be located at random, or at multiple unique sites. This has been concluded from the analysis of the distribution of gaps created by hydrolysis of the RNA segments with alkali with respect to the single Eco RI cleavage site (*Sugino, Y.* et al., 1975). The RNA segments, at least in the heavy strand, may be the remnants of the RNA primers of nascent fragments that have not been removed, possibly due to the conditions created by the inhibition of protein synthesis by chloramphenicol during their synthesis (*Helinski*, 1976). It is puzzling, though, that only a single RNA segment seems to persist per each circle and that the segments having such a distinct sequence are obviously located at many different sites in the heavy strand. If the RNA segment associated with the light strand is the whole or part of the primer RNA that initiated Col E1 DNA synthesis at the origin of replication, it should be complementary to the heavy-strand DNA at or near the origin of replication. Recently the sequence of the DNA around the origin of replication has been determined (*Tomizawa* et al., 1977; *Bastia,* 1977), but no sequence complementary to the light-strand RNA segment was found in this region. Therefore it seems as if the questions concerning the function of the RNA segments in Col E1 DNA synthesized in the presence of chloramphenicol and their possible role as primers for DNA synthesis are still open and have to await further studies to obtain clarification.

Much direct and indirect evidence strongly suggests RNA priming of the M13 or fd SS to RF conversion. The unambiguous identification of the primer RNA synthesized in vitro in crude cell extracts and with purified components has been confounded because other RNA was synthesized as well that was not covalently linked to DNA and therefore did not seem to serve primer functions. The size of RNA segments postulated to be covalently linked to newly synthesized DNA was estimated to be less than 50 nucleotides (*Geider* and *Kornberg,* 1974).

The structure of an RNA transcribed from the ori region of bacteriophage fd that is thought to be a primer RNA was recently determined by a different approach. It has been possible to isolate a hairpin-structured fragment of the viral DNA because it is protected against nuclease digestion by RNA polymerase, which specifically and strongly binds to it in the presence of DNA unwinding protein. This DNA region is an efficient promoter and appears to comprise the region where the primer RNA is being synthesized. This oriDNA, as it has been termed, consists of about 125 nucleotides and was found, by annealing it to restriction fragments produced by the endonuclease Hpa II, to map in the fragment H (*Schaller* et al., 1976). The origin of in vitro SS to RF conversion of M13 phage has been located in or near the same restriction fragment Hpa-H (*Tabak* et al., 1974). Through recent advances in DNA sequencing techniques (*Maxam* and *Gilbert,* 1977), it was possible to sequence the oriDNA (*Gray* et al., 1977). A corresponding RNA, called oriRNA, was synthesized on phage fd DNA complexed with *E. coli* DNA-binding protein by the action of *E. coli* RNA polymerase (*Geider* et al., 1978). This RNA has a length of about 30 nucleotides, and it gave a simple fingerprint after T1-nuclease digestion. The characterization of the digest products by fingerprinting and the analysis of the restriction pattern obtained after the RNA had been extended by DNA polymerase I yielded its localization in the origin region. The total sequence of the fd-oriRNA was deduced from the known DNA sequence in this region. The results suggest that the oriRNA starts at the end of a sequence that is protected by RNA polymerase. Six nucleotides upstream from the start point of the oriRNA, the DNA forms a hairpin structure. Half of this hairpin is transcribed, but then the RNA polymerase is presumably stopped by interfering DNA binding protein. Since binding of the DNA binding protein is not sequence specific, termination by this mechanism may cause the different 3′-OH ends of the isolated oriRNA.

RNA polymerase is also involved in the replication of λ DNA. In wild-type λ phage it presumably transcribes a small 4 S RNA, called oopRNA, from within the ori region of the λ DNA, which is believed to serve as primer to initiate replication (*Hayes* and *Szybalski*, 1973). The sequence of the oopRNA has been determined by RNA sequencing (*Dahlberg* and *Blattner*, 1973) and has been confirmed through the corresponding DNA sequence (*Scherer* et al., 1977). It is 81 nucleotides long and is the only λ RNA known that starts with pppG... However, whether it serves as primer has not been established unequivocally. Experiments aimed to show that it is covalently linked to newly synthesized DNA did not yield conclusive evidence (*Hayes* and *Szybalski,* 1975).

The synthesis of an RNA primer for the initiation of complementary strand synthesis of the replicative form (RF II) of bacteriophage G4 has been studied in great detail. This reaction requires three purified proteins: the *dna*G protein, the DNA-unwinding protein, and the DNA polymerase III holoenzyme complex (*Zechel* et al., 1975). The *dna*G protein has been shown to act as a rifampicin-resistant RNA polymerase on the DNA-unwinding protein-covered G4-single-stranded DNA. In the presence of ATP, GTP, and UTP several short RNA transcripts were synthesized. These segments can be isolated together with the template in the voided fraction of a gel filtration column (*Bouché* et al., 1975a). Gel filtration analysis of the RNA transcripts after their removal from the template showed that they were fairly homogeneous in size, consisting of about 20 nucleotides (*Bouché* et al., 1975b). Only one of these transcripts, though, is used as a primer for extention by DNA polymerase III holoenzyme and can be isolated associated with the 5′ end of the newly synthesized complementary DNA strand (*Bouché* et al., 1975a). This primer was sequenced and its provisional sequence, with some uncertainties in its 3′ region, was reported to be pppA-G-U-A-G-G-G-A-C-G-G-C-G-G-C-U-U-U-C-G-C-C-G-U-C-C-A-U...DNA (*Rowen* and *Bouché,* 1976; *Bouché* et al., 1978) A G-C-rich region in the middle part of the sequence can be aligned to form a hairpin-like secondary structure. The complementary sequence in the viral DNA probably also exists in a hairpin-like structure and serves as a promoter-like signal to which the *dna*G protein binds. The triphosphate group at the 5′ end indicates that this primer RNA survives its isolation in intact form.

The synthesis of polyoma nascent fragments in isolated nuclei from infected mouse 3T6 cells has been reported to be primed by an oligoribonucleotide called initiator RNA (iRNA) whose size was estimated from gel electrophoresis to be a decanucleotide. The iRNA starts with either ATP or GTP and has an intact triphosphate group at its 5′ end (*Eliasson* et al., 1974; *Reichard* et al., 1974). The sequence of the iRNA seems to be variable but its size seems to be constant. An iRNA, 9–11 ribonucleotides long, with a triphosphate group at the 5′ end and two or three deoxynucleotide residues at the 3′ end that are not removed by DNAse has also been found covalently linked to the nascent pieces synthesized in an in vitro system from cultured human lymphocytes (*Tseng* and *Goulian*, 1977).

As mentioned earlier, certain cellular transfer RNAs in eukaryotic cells may serve a dual purpose: as isoaccepting tRNA and as primer for the DNA synthesis on the genomes of RNA tumor viruses.

RNA tumor viruses are replicated via a DNA intermediate, called a provirus, which is synthesized on the viral 70 S RNA genome (*Temin, 1974*). Proviral DNA can be synthesized in vitro either with the endogenous RNA dependent-

DNA polymerase (reverse transcriptase) present in the virion (*Temin* and *Mitzutani,* 1970; *Baltimore,* 1970) or in reconstituted systems with isolated RNA and purified reverse transcriptase (*Faras* et al., 1973b; *Junghans* et al., 1975; *Haseltine* et al., 1976). In reconstituted systems, the enzyme isolated from Rous sarcoma virus accepts DNA- and RNA-DNA hybrids as templates for DNA synthesis and extends 3'-OH termini in a repair-like mode (*Duesberg* et al., 1971; *Hurwitz* and *Leis,* 1972; *Leis* and *Hurwitz,* 1972). The 70 S RNA complex extracted from purified virus has been found to be an excellent primer-template. This complex consists of two molecules of 35 S RNA (*King,* 1976) and several molecules of 4 S tRNA and 5 S rRNA (*Faras* et al., 1973a). The 35 S subunits are identical (*Beemon* et al., 1974; *Billeter* et al., 1974; *Quade* et al., 1974). When the RNAs were denatured by heat, and separated, the single species were found to be poor templates. However, they regained as much as 60% of the original template efficiency when allowed to reassociate (*Canaani* and *Duesberg,* 1972). Especially the reassociation of the 35 S species and the 4 S RNA seemed to be required, and it was concluded that a 4 S RNA hydrogen-bonded to the 35 S RNA may act as primer. The primer function of the 4 S RNA was shown by the isolation of that RNA covalently linked to newly synthesized DNA (*Verma* et al., 1971) and by demonstrating that $[\alpha\text{-}^{32}\text{P}]$ label was transferred from dC to rC and from dA to rA, respectively (*Flügel* et al., 1973). Denaturation and degradation of the DNA by pancreatic DNAse was shown to release an RNA species with the mobility in gels of tRNA (*Faras* et al., 1973a). In the case of AMV this 4 S RNA was identified as tryptophanyl tRNA (*Harada* et al., 1975). The tRNA^{Trp} found as primer RNA associated with the virions can be assumed to be identical with the normal cellular tRNA^{Trp} since in chicken cells there is only one isoaccepting tryptophanyl tRNA. The reason for its preferential use as primer for viruses of the avian sarcoma-leukosis group is not yet known. The association of the tRNA^{Trp} that can serve as primer seems to be very tight. Partial denaturation at 60 °C of the 70 S RNA complex, which would suffice to split the 35 S subunits and release all the loosely bound 5 S rRNAs and 4 S tRNA (including some tRNA^{Trp} that does not serve primer purposes), is apparently inadequate to remove the primer tRNA^{Trp}. The 35 S species isolated after the partial denaturation are still excellent template-primers in vitro. Upon further heating to 80 C, though, release of about one molecule tRNA^{Trp} per 35 S RNA and a concomitant decrease in template efficiency were observed (*Dahlberg* et al., 1974; *Sawyer* and *Dahlberg,* 1973; *Canaani* and *Duesberg,* 1972).

Another line of evidence suggested that the primer tRNA^{Trp} associated with the 70 S viral RNA is identical with the cellular tRNA^{Trp}: The template activity of isolated inactive 35 S RNA could be almost completely restored when tRNA^{Trp} isolated from uninfected cells was reannealed to the template 35 S RNA (*Faras* and *Dibble,* 1975; *Sawyer* et al., 1974; *Taylor, J.M.* et al., 1975). The tRNA^{Trp} appears to bind at a distinct site near the 5' terminus of the 35 S RNA (*Taylor* and *Illmensee,* 1975; *Cashion* et al., 1976; *Shine* et al., 1977). There is evidence that only part of the tRNA^{Trp} may be involved in the actual binding to the template. Nucleotides 2–17, as numbered from the 3' end, are resistant to RNAse attack in high salt, which may be an indication that they are tightly base-paired with the template (*Cordell* et al., 1976; *Eiden* et al., 1976). For efficient primer function, the whole, intact tRNA does not seem to be required. The 3′ half of the molecule, which can be obtained by controlled nucleolytic cleavage, still binds to the 35 S RNA at the 5′-terminal site and is recognized as primer terminus by the RNA-dependent DNA polymerase (*Brown* and *Armentrout, 1977*). Although, among the avian sarcoma leukosis virus group only tRNA^{Trp} seems to serve as primer, this tRNA^{Trp} needs not be from homologous sources. tRNA^{Trp} isolated from bovine liver has been shown to substitute for the avian-derived species and primes DNA synthesis on the 35 S avian RNA subunit (*Baroudy* et al., 1977). Whether bovine tRNA^{Trp} can functionally fully substitute for avian tRNA^{Trp}, though, is not completely clear, since the product of the DNA synthesis primed by the heterologous primer RNA has not been characterized.

V. Proteins Involved in RNA Primer Formation

A. RNA Polymerase of *E. coli*

As indicated by the rifampicin sensitivity of the DNA synthesis in several bacterial systems, the classical DNA-dependent RNA polymerase appears to be one of the enzymes that is involved in the synthesis of primer RNA. It is not yet clear how the primer-synthesis by RNA polymerase is regulated. Possibly, unique structures may exist, or may be shaped through the association of DNA-binding proteins on the DNA at or near the origin of replication as signals for RNA polymerase to bind and synthesize a primer RNA. One may expect similar structural elements in the DNA sequences around the origin of replication of, for instance, *E. coli*, bacteriophages, and plasmids whose replication depends on RNA polymerase, but these structural elements need not necessarily be similar to the promoter sequences known to be recognized by RNA polymerase for messenger RNA synthesis (*Gilbert, 1976*). RNA primers appear to be shorter than mRNA or even tRNA. Presumably, a termination signal for RNA polymerase follows shortly after the initiation signal. Whether termination requires accessory protein factors is not known, but for the regulation and timing of the primer-synthesis during the replicative cycle additional proteins are probably required. Analysis of temperature-sensitive mutants has shown that besides RNA polymerase at least the products of the genes *dna*A and *dna*C (*Wechsler* and *Gros*, 1971; *Zyskind* et al., 1977), of *dna*I (*Beyersmann* et al., 1974), and *dna*-252, which is a *dna*B mutant (*Zyskind* and *Smith*, 1977), may be involved in the initiation of DNA synthesis at the origin of *E. coli* replication. The *dna*H mutation, which has been described by *Sakai* et al. (1974) to cause a defect in the initiation at the nonpermissive temperature, probably does not code for still another protein. It appears to be a double mutant whose defect in initiation is due to a lesion in the *dna*A gene (*Derstine* and *Dumas*, 1976). The precise role of all these accessory proteins is unknown. However, available evidence seems to suggest that the *dna*A gene product functions like a repressor regulating the initiation of new rounds of chromosome replication in *E. coli* (*Messer* et al., 1975). RNA polymerase is apparently also responsible for the primer-synthesis in the conversion of viral strands of filamentous bacteriophages to the duplex replicative form II. Evidence obtained in vivo (*Horiuchi* and *Zinder,* 1976) and in vitro in cell-free extracts of *E. coli* (*Tabak* et al., 1974) suggests that the primer RNA is synthesized at a unique site on the genome of filamentous phages. Studies with purified components indicate that covering of the single-stranded template with DNA-unwinding protein is required to restrict the initiation of RNA synthesis by RNA polymerase to a single site (*Geider* and *Kornberg,* 1974). The presence of DNA-binding protein obviously directs RNA polymerase to start an RNA chain only at a unique initiation site on the viral DNA. RNA polymerase can initiate in the absence of DNA-binding protein at multiple sites on purified viral DNA from bacteriophage fl (*Stavriano-poulos* et al., 1972) or M13 (*Geider* et al., 1978). The unique site of the primer-synthesis appears to be part or all of the hairpin region in the viral DNA, which, probably because of its double-stranded structure, is not covered by single-strand (specific DNA-binding protein, and therefore remains accessible to RNA polymerase (*Schaller* et al., 1976). The origin for the synthesis of M13 progeny viral strands (*Suggs* and *Ray,* 1977) seems to be in the same region, but on the opposite strand. However, it remains to be seen whether this region, again, is a site where RNA polymerase initiates a primer RNA for viral DNA synthesis. Although the DNA-binding protein prevents RNA polymerase from forming primers adventiously on M13 (fd) DNA, it does not do so on φX174 viral DNA. The in vitro system that primes M13 viral strand will also prime φX174 viral strands, and these primers can be utilized by DNA polymerase I or III for the synthesis of φX174 complementary DNA in vitro. This reaction may be a salvage pathway for the priming of φX174 DNA replication in vivo when the usual priming mechanism is inoperative. For instance, the φX174 replication in a *dna*C mutant at the nonpermissive temperature may be primed by RNA polymerase (*Dumas* et al., 1975). However, in normal *E. coli* cells, the parental RF II synthesis of φX174 is, in crude extracts in vivo and in vitro, completely insensitive to rifampicin and therefore independent of RNA polymerase. To explain the distinction between both single-strand phage DNAs, it has been suggested that factors exist that allow primer-synthesis by RNA polymerase only on M13 (fd) DNA, but not on φX174 DNA. One such discriminatory factor has been described that forms a complex with the RNA polymerase. It appears to be a protein that is released from the enzyme by rifampicin (*Wickner, W.* and *Kornberg,* 1974), and its presence directs RNA polymerase to synthesize a primer RNA only on M13 DNA and to ignore φX174 DNA in vitro.

Two other factors described by *Vicuna* et al. (1977a, b) operate by a different mechanism. They appear to protect specifically the primer formed on fd DNA from being degraded by RNAse H, a constituent of the in vitro assay mixture. RNA that may have been formed erroneously on φX174 DNA seems to be eliminated by the enzyme before it can serve as primer for DNA synthesis.

B. DNA Primases

The term DNA primase was first suggested for the T7 gene 4 protein (*Scherzinger* et al., 1977). It was probably meant to imply that this protein, although in

principle an RNA (or DNA) polymerase, is highly specialized for the synthesis
of primers for DNA replication. This definition would also apply for the *dnaG*
protein. The gene 41 protein of bacteriophage T4 (*Alberts* et al., 1975) may
also turn out to be a DNA primase.

1. T7 Gene 4 Protein

Genetic and biochemical analysis have shown that two proteins, the phage-
induced DNA polymerase and the protein encoded by gene 4 of phage T7,
play a major role in the replication of bacteriophage T7. The T7 DNA polymerase, as all other known DNA polymerases, is unable to initiate DNA chains
de novo. However, in combination with gene 4 protein, extensive DNA synthesis
was catalyzed in vitro by T7 DNA polymerase on the natural template, double-
stranded T7 DNA (*Strätling* et al., 1973; *Hinkle* and *Richardson,* 1974) as well
as on a variety of other templates (*Scherzinger* and *Litfin,* 1974). Stimulation
of DNA synthesis by the gene 4 protein seemed to suggest a function of this
protein in the initiation of DNA strands (*Hinkle* and *Richardson,* 1974; *Scherzinger* and *Klotz,* 1975), and the stimulation of the reaction by ribonucleoside
triphosphates seemed to indicate that an RNA might be synthesized to serve
as primer. Neither enzyme alone showed RNA polymerase activity, yet, together,
purified gene 4 protein and T7 DNA polymerase synthesize in vitro, in a ribonucleoside triphosphate-dependent reaction on single-stranded φX174 DNA, an
RNA-DNA copolymer whose RNA part consists of a tetranucleotide of the
sequence pppApCpCpA-(DNA).. (*Scherzinger* et al., 1977a, b). However, the
appearance of the primer by which DNA synthesis on duplex T7 DNA is
initiated, and whether it is an RNA, are not yet known. The ribonucleoside
triphosphates that stimulate the replication of T7 DNA two to fourfold may
also be required for another function that has recently been suggested for the
T7 gene 4 protein (*Scherzinger* et al., 1977a; *Kolodner* and *Richardson,* 1977).
The T7 gene 4 protein has been shown to be a single-strand DNA-dependent
nucleoside 5'-triphosphatase, and it appears also to hydrolyze triphosphates
during the replication of the duplex T7 DNA template. It was proposed that
the hydrolysis serves to provide the energy for the necessary unwinding of
the DNA strands. With respect to this function, the T7 gene 4 protein may
be comparable to the rep-protein of *E. coli* (*Scott* et al., 1977).
Highly purified T7 gene 4 protein preparations consist of a mixture of
two polypeptides with molecular weights of 57,000 and 66,000 (*Hinkle* and *Richardson,* 1975; *Scherzinger* et al., 1977a). Both polypeptides seem to be the
product of the T7 gene 4 because both are absent in cells infected with T7
gene 4 amber mutants. Since the tryptic peptide maps of the two polypeptides
labeled with radioactive ^{125}I are almost identical, the small polypeptide could
be a proteolytic cleavage product of the larger one. Looking at DNA polymerase
I of *E. coli* as an example of the fact that proteolytic cleavage of a multifunctional
protein can lead to the separation of enzymatic activities (*Kornberg,* 1974),
one might wonder whether the two polypeptides found in the T7 gene 4 protein
preparations may also differ in functional capabilities.

2. The *dna*G Protein

Synthesis of the parental replicative form of the bacteriophages φX174 and G4, a close relative of φX174 (*Godson,* 1974), is not inhibited by rifampicin, nor is the initiation of the nascent fragments during the replication of the *E. coli* chromosomal DNA inhibited by the drug. Yet, evidence, which has been discussed above, suggests that in all three cases DNA synthesis is primed by RNA. The synthesis of such a primer RNA clearly would require a rifampicin-resistant RNA polymerase. In vitro studies of the parental RF II synthesis of bacteriophage G4 have yielded evidence that the *dna*G protein may possess the rifampicin-resistant RNA polymerase activity that is required for the synthesis of the G4 RNA primer. The *dna*G protein catalyzes the polymerization of all four ribonucleoside triphosphates on G4 viral strands covered with DNA-binding protein. This RNA synthesis is not affected by rifampicin. Although in the in vitro reaction several short oligoribonucleotides are generated, apparently only one serves as primer for the synthesis of the complementary DNA strand (*Bouché* et al., 1975 b). The others are presumably degraded, or removed during the elongation of the DNA chain by a strand displacement reaction. The *dna*G protein was originally characterized by a temperature-sensitive mutant of *E. coli,* which, upon transfer to the nonpermissive temperature, would cease DNA synthesis almost immediately. On the basis of this result it was proposed that the *dna*G gene product is involved in DNA chain elongation (for review see *Gross,* 1972). Closer examination of this phenomenon in *E. coli* cells led *Lark* (1972 a) to conclude that the *dna*G protein may participate in the priming of the Okazaki pieces. Studies of phage growth in *dna*G ts mutants at the nonpermissive temperature have shown that a functional *dna*G gene product is also required during the initial stages of the reproductive cycle of φX174 (*Mc Fadden* and *Denhardt,* 1974) and Ha$_2$, a φX174 mutant able to grow on *E. coli* K12 (*Truffaut* and *Manheimer,* 1975). The *dna*G function is not required, for the parental RF formation from the infecting M13 viral strand, but it seems necessary for the RF to RF synthesis step (*Ray* et al., 1975). Surprisingly, the parental RF formation of phage G4, which was shown to be dependent on the *dna*G protein in vitro, did not appear to be impaired in vivo at the nonpermissive temperature in two *dna*G ts mutants (*Derstine* et al., 1976). However, the subsequent step, the RF to RF replication, would not take place in these mutants at the nonpermissive temperature. To reconcile the discrepancy between in vitro and in vivo results the authors speculate that the *dna*G protein coded by the alleles they used in their experiments may have a partial lesion that would affect its function during RF to RF synthesis but not its function during the priming of the parental RF synthesis.

A functional *dna*G protein also seems to be necessary in several other systems. Thus, the synthesis of the putative primer RNA (oopRNA) for bacteriophage λ replication depends on a functional *dna*G protein. Besides the *dna*G protein, the *dna*B protein, RNA polymerase, and the products of the λ genes O and P are required (*Hayes* and *Szybalski,* 1975). The replication of plasmid Col E1 DNA was shown to be dependent on the *dna*G function (*Collins* et al., 1975). The bacteriophage P2 also requires a functional *dna*G gene product for its replication, but the satellite phage P4 seems to be replicated by a different mechanism. Its reproduction is normal in *dna*G ts mutants at the nonpermissive temperature (*Bowden* et al., 1975). Perhaps the rifampicin-resistant RNA polymerase activity induced by P4 can substitute for the *dna*G function for the primer synthesis (*Barrett* et al., 1972).

The *dna*G gene has been mapped between the tol C and the uxa C marker genes (*Chen* and *Carl,* 1975) corresponding to a location at about 60 min on the recalibrated *E. coli* linkage map *(Bachmann* et al., 1976). The availability of the *dna*G ts mutants has greatly facilitated the isolation of the *dna*G protein. They provided the basis for the development of a complementation assay that was used to monitor the purification procedure. Three laboratories reported the isolation of the *dna*G protein (*Klein* et al., 1973; *Wickner, S.* et al., 1973; *Bouché* et al., 1975b; *Rowen* and *Kornberg,* 1978a). All three procedures yield a protein of approximately 60,000 daltons. According to sedimentation studies and gel electrophoresis under native and denaturing conditions, it appears to exist in solution as a monomeric molecule. When the protein is purified from *dna*G ts cells, it is more thermolabile than the wild-type enzyme, which is strong evidence that the purified protein is indeed the *dna*G gene product. As mentioned earlier, the *dna*G protein exhibits rifampicin-resistant RNA polymerase activity on DNA-binding protein-covered G4 viral strands if all four ribonucleoside triphosphates are present. It does not accept M13 viral strand in the presence of DNA unwinding protein (*Schekman* et al., 1974). The resolution of the parental RF formation of φX174 in vitro into a two-stage reaction showed that the *dna*G protein is not required during the first stage when the viral DNA template is converted into a so-called activated form. The *dna*G protein is required, presumably for the synthesis of the primer on this activated form DNA, during the second stage (*Ray* et al., 1976). This activated form DNA is probably identical to an isolatable "replication intermediate" that is formed upon incubation of five proteins—protein i, protein n, DNA-unwinding protein, *dna*B protein, and *dna*C protein—with ATP and φX174 viral DNA (*Weiner* et al., 1976). In the replication intermediate the *dna*B protein has been suggested to function as a "mobile promoter" serving as a recognition signal for the *dna*G protein (*McMacken* et al., 1977). When *dna*G protein is incubated with the replication intermediate and the four ribonucleoside triphosphates, multiple short RNA transcripts, each approximately 20 nucleotides long, are synthesized (*McMacken* and *Weiner,* 1976). Recently, however, it was found that the *dna*G protein not only has RNA polymerase activity forming oligonucleotides from ribonucleoside triphosphates but it also seems to have DNA polymerase activity, and can also incorporate deoxyribonucleoside triphosphates into short oligodeoxyribonucleotide chains on the G4 template in vitro (*Kornberg,* 1977). Even a hybrid ribonucleotide-deoxyribonucleotide transcript is formed if both ribonucleotide triphosphates and deoxyribonucleotide triphosphates are present in the reaction mixture (*Rowen* and *Kornberg,* 1978b). Either nucleotide can serve as primer for chain elongation by DNA polymerase II or III in combination with *dna*Z protein and DNA elongation factors I and III. The product is an almost full size complementary strand (*Wickner, S.,* 1977). It remains to be seen which of the priming modes is physiologically relevant.

Perhaps both RNA and DNA priming by the *dna*G protein, can occur in parallel inside the cell. Then the *dna*G protein would be the first example of a DNA polymerase capable of initiating a new chain.

Besides the *dna*G protein several other rifampicin-resistant RNA polymerases have been described. Most of them are either products of bacteriophage genes, like the RNA polymerase of T7 (*Chamberlin* et al., 1970) and T3 (*Dunn* et al., 1971; *Maitra,* 1971), or at least they are induced after infection, as is the case with the *Bacillus subtilis* phage PBS2 (*Clark* et al., 1974) or the *Pseudomonas putida* phage gh-1 (*Towle* et al., 1975). A rifampicin-resistant RNA polymerase has also been found associated with the virion of the coliphage N4 (*Pesce* et al., 1976). However, whether these RNA polymerases play a role in primer synthesis during phage DNA replication remains to be shown.

Another rifampicin-resistant RNA polymerase activity present in cytoplasmic membranes and DNA protein complexes isolated from uninfected *E. coli* cells has been purified and tested for its possible involvement in primer-RNA synthesis for DNA replication, but no positive evidence was obtained (*Ohasa* and *Tsugita,* 1976). This enzyme is structurally very different from the *dna*G protein. The properties of the enzyme isolated from a *dna*G ts mutant are not different from those of the enzyme isolated from wild-type cells. It does not substitute for the *dna*G protein in priming G4 complementary strand synthesis in vitro nor does it complement *dna*G ts mutant extracts in vitro. Therefore it is apparently not related to the *dna*G protein.

VI. Excision of the RNA Primer

The necessarily transient nature of RNA primers calls for enzymatic mechanisms for the excision of the primer RNA. In *E. coli* the 5′→3′ exonucleolytic activity of DNA polymerase(s) seems to be the most likely candidate for this job, although the specificity of ribonuclease H for the RNA part of an RNA-DNA hybrid may also be perfectly suited for the removal of an RNA primer.

A. The 5′→3′ Exonuclease Activity of DNA Polymerase I

DNA polymerase I of *E. coli* has, besides its polymerase activity also a 5′→3′ exonuclease activity capable of degrading polyribonucleotides and polydeoxyribonucleotides (*Kornberg,* 1974; *Lehman* and *Uyemura,* 1976). In vitro, this nuclease function was shown to remove the oligoribonucleotide primers from the 5′ end of a growing DNA chain (*Roychoudhury* and *Kössel,* 1973). Both functions of DNA polymerase I are required in a concerted action together with *E. coli* ligase to convert the in vitro-synthesized M13 replicative form II into the covalently closed double-stranded replicative form I. The polymerase is believed to fill in the gap in the complementary strand of the RF II while the 5′→3′ nuclease excises the RNA primer. The two adjacent DNA ends can then be joined by *E. coli* ligase (*Geider* and *Kornberg,* 1974). Interestingly, the DNA polymerase III holoenzyme, which is present for the synthesis of the complementary strand of RF II, is obviously unable to fill the gap and/or excise the RNA-priming fragment, although it too has a $5' \rightarrow 3'$ exonuclease associated with it (*Livingston* and *Richardson*, 1975). It is not clear whether the $5' \rightarrow 3'$ exonuclease of DNA polymerase I is also responsible for the primer excision in vivo. Col E1 DNA isolated from a thermosensitive pol A$^-$ mutant after a shift to the nonpermissive temperature is sensitive to alkali and pancreatic RNAse. It has been suggested that this may be due to the lack of an excising nuclease function (*Goebel* and *Schrempf*, 1973). However, this evidence seems rather indirect. More direct evidence comes from the studies on a conditional lethal mutant of *E. coli* defective only in the $5' \rightarrow 3'$ exonuclease function of DNA polymerase I. These cells accumulate small nascent DNA fragments at the nonpermissive temperature. Since the polymerase I activity is normal under these conditions, the accumulation of short fragments seems to be due to the missing nuclease function, which may be required to remove RNA primers from those fragments (*Konrad* and *Lehman*, 1974). Without excision of the RNA *E. coli* ligase cannot seal the nicks between fragments because *E. coli* ligase is unable to join RNA and DNA.

B. Ribonuclease H

RNAse H was first isolated from calf thymus (*Hausen* and *Stein*, 1970), but it has also been found in *E. coli* (*Henry* et al., 1973; *Miller* et al., 1973; *Berkower* et al., 1973), in yeast (*Wyers* et al., 1973), and associated with the RNA-dependent DNA polymerase of RNA tumor viruses (*Mölling* et al., 1971).

RNAse H specifically degrades the RNA moiety of RNA-DNA hybrids and therefore seems especially suited for the excision of RNA primers. The RNAse H activity found as an unseparable constituent of the viral reverse transcriptase (*Keller* and *Crouch*, 1972; *Baltimore* and *Smoler*, 1972; *Leis* et al., 1973) may be conceived to have an obvious function in the degradation of the tRNA primer and eventually of the whole viral RNA to allow synthesis of a complete double-stranded proviral DNA.

Whether the cellular RNAse H is engaged in the excision of RNA primers remains to be demonstrated. In *E. coli* it is probably not. One could argue that, in *E. coli*, RNAse H cannot substitute for the $5' \rightarrow 3'$ nuclease function of DNA polymerase I. If it were able to substitute, one should expect to see no accumulation of Okazaki fragments in the $5' \rightarrow 3'$ nuclease-deficient mutant of *Konrad* and *Lehman* (1974) at the nonpermissive temperature, because RNAse H, which was probably not defective in this mutant, could have removed the primer RNA thought to be obstructive for the joining of the fragments.

VII. Conclusions

There can be no doubt that, in a number of cases, RNA serves as a primer for DNA synthesis. Ample demonstration in vitro that RNA can be a primer for DNA chain initiation by DNA polymerase from prokaryotic as well as eukaryotic organisms may justify the conclusion that RNA priming may also be one way in which DNA synthesis is initiated in vivo. The inhibition of the replication in bacterial systems in vivo by rifampicin is one major argument for this conclusion. Primer-synthesis by RNA polymerase, which is affected by the drug, may be especially advantageous for the initiation of DNA synthesis at specific sites, for example at the origin of replication. RNA polymerase of *E. coli* is, in contrast to DNA polymerase, able to recognize specific signals on DNA. It can, in addition, be directed by factors to respond to such signals at a certain time during the replicative cycle. Therefore, the initiation of new rounds of DNA replication can be effectively controlled via regulation of primer-RNA synthesis by RNA polymerase. It can be hoped that cloned fragments of the *E. coli* genome containing the origin of replication, like the one described by *Marsh* and *Worcel* (1977), will facilitate the studies on the regulation of the DNA chain initiation at the origin of replication and will also facilitate the determination of the components that are involved in this step.

For the more basic questions about the mechanisms of primer-RNA synthesis and its degradation, the parental RF II formation of the filamentous bacteriophages M13, fd, f1 still appears to provide the most promising system because of its relative simplicity and its advanced characterization. It also may offer the best possibilities at present to correlate in vitro and in vivo data. Insofar as the initiation site for the primer synthesis appears to be identical in vivo and in vitro there is a high probability that results on the primer-RNA synthesis obtained in vitro indeed reflect the in vivo situation. Furthermore, because the part of the sequence of the fd genome containing the replication origin is known (*Gray* et al., 1977), and because it was possible to pinpoint exactly the base sequence where RNA polymerase initiates and terminates the primer RNA in vitro it may also become possible to elucidate the fate of that primer inside the cell.

Another reason to claim tht RNA is priming DNA synthesis in vivo is the observation that RNA fragments, covalently linked to short nascent DNA (Okazaki pieces), could be isolated from pulse-labeled *E. coli* cells. However, the apparent difficulties involved in repeating these experiments together with the discovery of an excision-repair mechanism that was shown to produce fragments similar in size to Okazaki fragments without involving RNA primers (*Tye* et al., 1977) make this argument crumble. Aside from these findings it looks as if nascent DNA pieces during discontinuous chain elongation could be, but *need not* be initiated by RNA. The recently discovered ability of the *dna*G protein to synthesize in vitro oligonucleotide primers on G4 DNA either from ribonucleoside triphosphates and/or deoxyribonucleoside triphosphates makes it necessary to consider the possibility of Okazaki pieces without RNA primers in bacterial systems in vivo.

Good evidence for RNA priming of discontinuous replication has been obtained in eukaryotic in vitro systems. However, due to the complexity of these systems almost nothing is known about the enzymes involved in the primer-RNA synthesis. The primers isolated from in vitro replication systems from animal cells were found to be six to ten ribonucleotides long. They have apparently no specific sequence. In contrast to earlier reports, the lack of sequence specificity has been recently confirmed for bacterial RNA primers of discontinuous replication. The question then arises as to how these primer RNAs are initiated and how they are terminated. Perhaps the suggestion by *Reichard* et al. (1974) is correct that the length of the primer is the signal for its termination and that its sequence is unimportant. But what is the signal for its start? No experimental proof has yet been provided for the theory put forward by *Okazaki* et al. (1973), i.e., that signals for the initiation may exist at regular intervals on the DNA. Based on observations made during bacteriophage fd SS to RF conversion in vitro, *Gefter* and *Sherman* (1977) suggest that self-complementary sequences may play an important role as signals for the initiation of Okazaki fragments. It appears as if one key to the solution of the recognition problem lies in the DNA sequence. Therefore it can be hoped that the examination of known and forthcoming DNA sequences will provide some clues as to whether such signals exist and what their structure might be.

Acknowledgments. I wish to thank Drs. K. *Geider*, G. *Hobom*, and W. *Staudenbauer* for contribution of results prior to publication and valuable constructive criticism, and I thank A. *Zechel* for help with the preparation of the manuscript.

References

Abdel-Monem, M., Chanal, M.C., Hoffmann-Berling, H.: DNA unwinding enzyme II of *Escherichia coli*. 1. Purification and characterization of the ATPase activity. Eur. J. Biochem. *79*, 33–38 (1977a)

Abdel-Monem, M., Dürwald, H., Hoffmann-Berling, H.: Enzymic unwinding of DNA. 2. Chain separation by an ATP-dependent DNA unwinding protein. Eur. J. Biochem. *65*, 441–449 (1976)

Abdel-Monem, M., Dürwald, H., Hoffmann-Berling, H.: DNA-unwinding enzyme II of *Escherichia coli*. 2. Characterization of the DNA unwinding activity. Eur. J. Biochem. *79*, 39–45 (1977b)

Alberts, B., Sternglanz, R.: Recent excitements in the DNA replication problem. Nature (London) *269*, 655–661 (1977)

Alberts, B., Barry, J., Bittner, M., Davies, M., Hama-Inaba, H., Liu, C.-C., Mace, D., Moran, L., Morris, C.F., Piperno, J., Sinha, K.: In vitro DNA replication catalyzed by six purified T4 bacteriophage proteins. In: Nucleic Acid-Protein Recognition, *Vogel, H.J.* (Ed.), New York: Academic Press, 1977, pp. 31–63

Alberts, B., Morris, C.F., Mace, D., Sinha, N., Bittner, M., Moran, L.: Reconstruction of the T4 bacteriophage replication apparatus from purified components. In: DNA Synthesis and its Regulation, *Goulian, M., Hanawalt, P., Fox, C.F.* (Eds.), Menlo Park, Calif.: W.A. Benjamin, Inc. 1975, pp. 241–268

Anderson, S., Kaufmann, G., Depamphilis, M.L.: RNA primers in SV40 DNA replication. I. Identification of transient RNA-DNA covalent linkages in replicating DNA. Biochemistry *16*, 4990–4998 (1977)

Bachmann, B.J., Low, K.B., Taylor, A.L.: Recalibrated linkage map of *Escherichia coli* K12. Bacteriol. Rev. *40*, 116–167 (1976)

Baltimore, D.: Viral RNA-dependent DNA polymerase. Nature (London) *226*, 1209–1211 (1970)

Baltimore, D., Smoler, D.: Association of an endoribonuclease with the avian myeloblastosis virus deoxyribonucleic acid polymerase. J. Biol. Chem. *247*, 7282–7287 (1972)

Baroudy, B.M., Fournier, M., Labouesse, J., Papas, T.S., Chirikjian, J.G.: Transfer RNA[Trp] (bovine) binding to reverse transcriptase of avian myeloblastosis virus and function as heterologous primer. Proc. Natl. Acad. Sci. *74*, 1889–1893 (1977)

Barrett, K.J., Gibbs, W., Calendar, R.: A transcribing activity induced by satellite phage P4. Proc. Natl. Acad. Sci. *69,* 2986–2990 (1972)

Bastia, D.: The nucleotide sequence surrounding the origin of DNA replication of Col E1. Nucl. Acid Res. *4,* 3123–3142 (1977)

Bazzicalupo, P., Tocchini-Valentini, G.P.: Curing of an *Escherichia coli* episome by rifampicin. Proc. Natl. Acad. Sci. *69,* 298–300 (1972)

Beemon, K., Duesberg, P., Vogt, P.: Evidence for crossing-over between avian tumor viruses based on analysis of viral RNAs. Proc. Natl. Acad. Sci. *71,* 4254–4258 (1974)

Berger, H., Huang, R.C.C.: Studies on nascent DNA in mouse myeloma. Cell *2,* 23–30 (1974)

Berkower, I., Leis, J., Hurwitz, J.: Isolation and characterization of an endonuclease from *Eschericia coli* specific for ribonucleic acid in ribonucleic acid-deoxyribonucleic acid hybrid structures. J. Biol. Chem. *248,* 5914–5921 (1973)

Bessman, M.J., Lehman, I.R., Adler, J., Zimmerman, S.B., Simms, E.S., Kornberg, A.: Enzymatic synthesis of deoxyribonucleic acid. III. The incorporation of pyrimidine and purine analogues into deoxyribonucleic acid. Proc. Natl. Acad. Sci. *44,* 633–640 (1958)

Beyersmann, D., Messer, W., Schlicht, M.: Mutants of *Eschericia coli* B/r defective in deoxyribonucleic acid initiation: *dna* I, a new gene for replication. J. Bacteriol.. *118,* 783–789 (1974)

Billeter, M.A., Parson, J.T., Coffin, J.M.: The nucleotide sequence complexity of avian tumor virus RNA. Proc. Natl. Acad. Sci. *71,* 3560–3564 (1974)

Bird, R.E., Louarn, J., Martuscelli, J., Caro, L.: Origin and sequence of chromosome replication in *Escherichia coli.* J. Mol. Biol. *70,* 549–566 (1972)

Blair, D.G., Clewell, D.B., Sheratt, D.J., Helinski, D.R.: Strand-specific supercoiled DNA-protein relaxation complexes: comparison of the complexes of bacterial plasmids Col E1 and Col E2. Proc. Natl. Acad. Sci. *68,* 210–214 (1971)

Blair, D.G., Sheratt, D.J., Clewell, D.B., Helinski, D.R.: Isolation of supercoiled colicinogenic factor E1 DNA sensitive to ribonuclease and alkali. Proc. Natl. Acad. Sci. *69,* 2518–2522 (1972)

Blinkerd, P.E., Toliver, A.P.: Association of RNA with discontinuous DNA-replication in HeLa-cells. Cytobios *10,* 221–233 (1974)

Bouché, J.P., Schekman, R., Weiner, J., Zechel, K., Kornberg, A.: Multienzyme systems in the replication of φX174 and G4 DNAs. In: DNA Synthesis and its Regulation, Goulian, M., Hanawalt, P., Fox, C.F. (Eds.), Menlo Park, Calif.: W.A. Benjamin, Inc. 1975a, pp. 210–226

Bouché, J.P., Zechel, K., Kornberg, A.: dnaG gene product, a rifampicin-resistant RNA polymerase, initiates the conversion of a single-stranded coliphage DNA to its duplex replicative form. J. Biol. Chem. *250,* 5995–6001 (1975b)

Bouché, J.-P., Rowen, L., Kornberg, A.: The RNA primer synthesized by primase to initiate phage G4 DNA replication. J. Biol. Chem. *253,* 765–769 (1978)

Bowden, D., Twersky, R., Calendar, R.: E. coli DNA synthesis mutants: Their effect on bacteriophage P2 and satellite bacteriophage P4 DNA synthesis. J. Bacteriol. *124,* 167–175 (1975)

Brown, R.D., Armentrout, R.W.: Primer recognition by avian-myeloblastiosis virus RNA-directed DNA-polymerase. J. Virol. *21,* 1236–1239 (1977)

Brutlag, D., Schekman, R., Kornberg, A.: A possible role for RNA polymerase in the initiation of M13 DNA synthesis. Proc. Natl. Acad. Sci. *68,* 2826–2829 (1971)

Buckley, P.J., Kosturko, L.D., Kozinski, A.W.: In vivo production of an RNA-DNA copolymer after infection of *Escherichia coli* by bacteriophage T4. Proc. Natl. Acad. Sci. *69,* 3165–3169 (1972)

Canaani, E., Duesberg, P.H.: Role of subunits of 60 to 70S avian tumor virus ribonucleic acid in the template activity for the viral deoxyribonucleic acid polymerase. J. Virol. *10,* 23–31 (1972)

Casey, J., Davidson, N.: Rates of formation and thermal stabilities of RNA:DNA and DNA:DNA duplexes at high concentrations of formamide. Nucl. Acid Res. *4,* 1539–1552 (1977)

Cashion, L.M., Joho, R.H., Planitz, M.A., Billeter, M.A., Weissmann, C.: Initiation sites of Rous-sarcoma virus RNA-directed DNA-synthesis in vitro. Nature (London) *262*, 186–190 (1976)

Cassani, G., Burgess, R.R., Goodman, H.M., Gold, L.: Inhibition of RNA polymerase by streptolydigin. Nature (London) New Biol. *230*, 197–200 (1971)

Chamberlin, M., McGrath, J., Waskell, L.: New RNA polymerase from *Escherichia coli* infected with bacteriophage T7. Nature (London) *228*, 227–231 (1970)

Chang, L.M.S., Bollum, F.J.: A chemical model for transcriptional initiation of DNA replication. Biochem. Biophys. Res. Commun. *46*, 1354–1360 (1972)

Chargaff, E.: Initiation of enzymic synthesis of deoxyribonucleic acid by ribonucleic acid primers. Progr. Nucleic Acid Res. Mol. Biol. *16*, 1–24 (1976)

Chen, P.L., Carl, P.L.: Genetic map location of the *Escherichia coli dna*G gene. J. Bacteriol. *124*, 1613–1614 (1975)

Clark, S., Losick, R., Pero, J.: New RNA polymerase from *Bacillus subtilis* infected with phage PBS2. Nature (London) *252*, 21–24 (1974)

Clewell, D.B., Evenchik, B.G.: Effects of rifampicin, streptolydigin and actinomycin-D on replication of Col E1 plasmid DNA in *Escherichia coli.* J. Mol. Biol. *75*, 503–513 (1973)

Clewell, D.B., Evenchik, B., Cranston, J.W.: Direct inhibition of Col E1 plasmid DNA replication in *Escherichia coli* by rifampicin. Nature (London) New Biol. *237*, 29–31 (1972)

Collins, J., Williams, P., Helinski, D.R.: Plasmid Col E1 DNA replication in *Escherichia coli* strains temperature-sensitive for DNA-replication. Mol. Gen. Genet. *136*, 273–290 (1975)

Cordell, B., Stavnezer, E., Friedrich, R., Bishop, J.M., Goodman, H.M.: Nucleotide-sequence that binds primer for DNA-synthesis to the avian sarcoma virus genome. J. Virol. *19*, 548–558 (1976)

Crawford, L.V., Robbins, A.K., Nicklin, P.M., Osborn, K.: Polyoma DNA-replication—Location of the origin in different virus strains. Cold Spring Harbor Symp. Quant. Biol. *39*, 219–225 (1974)

Crippa, M., Tocchini-Valentini, G.: Synthesis of amplified DNA that codes for ribosomal RNA. Proc. Natl. Acad. Sci. *68*, 2769–2773 (1971)

Dahlberg, J.E.: RNA primers for the reverse transcriptase of RNA tumor viruses. In: Nucleic Acid-Protein Recognition, *Vogel, H.J.* (Ed.), New York: Academic Press, 1977, pp. 345–358

Dahlberg, J.E., Blattner, F.R.: Sequence of a self-terminating RNA made near the origin of DNA replication of phage λ. Fed. Proc. *32*, 664Abs (1973)

Dahlberg, J.E., Sawyer, R.C., Taylor, J.M., Faras, A.J., Levinson, W.E., Goodman, H.M., Bishop, J.M.: Transcription of DNA from 70S RNA of Rous-sarcoma virus. 1. Identification of a specific 4S RNA which serves as primer. J. Virol. *13*, 1126–1133 (1974)

Dasgupta, S.: The role of primer RNA in coliphage M13 replicative form DNA replication. Fed. Proc. *36*, 847–847 (1977)

Denhardt, D.T.: The isometric single stranded DNA phages. In: Comprehensive Virology, Vol. 7: Reproduction: Bacterial Viruses, *Fraenkel-Conrat, H., Wagner, R.R.* (Eds.), New York: Plenum Publishing Comp. 1977, pp. 1–104

Derstine, P.L., Dumas, L.B.: Deoxyribonucleic acid synthesis in a temperature-sensitive *Escherichia coli dna*H mutant strain HF 4704S. J. Bacteriol. *128*, 801–809 (1976)

Derstine, P.L., Dumas, L.B., Miller, C.A.: Bacteriophage G4 DNA synthesis in temperature-sensitive mutants of *Escherichia coli.* J. Virol. *19*, 915–924 (1976)

Dressler, D., Wolfson, J., Magazin, M.: T7 DNA-replication. 2. Initiation and reinitiation of DNA-synthesis during replication of bacteriophage T7. Proc. Natl. Acad. Sci. *69*, 998–1002 (1972)

Dove, W.F., Inokuchi, H., Stevens, W.F.: Replication control in phage lambda. In: The Bacteriophage Lambda, *Hershey, A.D.* (Ed.), Cold Spring Harbor: Cold Spring Harbor Laboratory, 1971, pp. 747–771

Duesberg, P., von der Helm, K., Canaani, E.: Comparative properties of RNA and DNA templates for the DNA polymerase of Rous sarcoma virus. Proc. Natl. Acad. Sci. *68*, 2505–2509 (1971)

Dumas, L.B., Miller, C.A., Bayne, M.L.: Rifampicin inhibition of bacteriophage φX174 parental replicative-form DNA synthesis in an *Escherichia coli dna*C mutant. J. Virol. *16*, 575–580 (1975)

Dunn, J.J., Bautz, F.A., Bautz, E.K.F.: Different template specificities of phage T3 and T7 RNA polymerases. Nature (London) New Biol. *230*, 94–96 (1971)

Edenberg, H.J., Huberman, J.A.: Eucaryotic chromosome replication. Ann. Rev. Genet. *9*, 245–284 (1975)

Eiden, J.J., Quade, K., Nichols, J.L.: Interaction of tRNA^{trp} with Rous sarcoma virus 35S RNA. Nature (London) *259*, 245–247 (1976)

Eisenberg, S., Griffith, J., Kornberg, A.: φX174 cistron A protein is a multifunctional enzyme in DNA replication. Proc. Natl. Acad. Sci. *74*, 3198–3202 (1977)

Eliasson, R., Martin, R., Reichard, P.: Characterization of RNA initiating discontinuous synthesis of polyoma DNA. Biochem. Biophys. Res. Commun. *59*, 307–313 (1974)

Ezekiel, D.H., Hutchins, J.E.: Mutations affecting RNA polymerase associated with rifamycin resistance in *E. coli*. Nature (London) *220*, 276–277 (1968)

Faras, A.J., Dibble, N.A.: RNA-directed DNA-synthesis by DNA-polymerase of Rous-sarcoma virus—structural and functional identification of 4S primer RNA in uninfected cells. Proc. Nat. Acad. Sci. *72*, 859–863 (1975)

Faras, A.J., Garapin, A.C., Levinson, W.E., Bishop, J.M., Goodman, H.M.: Characterization of the low-molecular-weight RNAs associated with the 70S RNA of Rous sarcoma virus. Virology *12*, 334–342 (1973a)

Faras, A.J., Taylor, J.M., Levinson, W.E., Goodman, H.M., Bishop, J.M.: RNA directed DNA polymerase of Rous sarcoma virus initiation of synthesis with 70S viral RNA as template. J. Mol. Biol. *79*, 163–183 (1973b)

Fareed, G.C., Salzman N.P.: Intermediates in SV 40 DNA chain growth. Nature (London) New Biol. *238*, 274–277 (1972)

Ficq, A., Brachet, J.: RNA-dependent DNA polymerase: Possible role in the amplification of ribosomal DNA in *Xenopus* oocytes. Proc. Natl. Acad. Sci. *68*, 2774–2776 (1971)

Fidanián, H.M., Ray, D.S.: Replication of bacteriophage M13. VII. Requirement of the gene 2 protein for the accumulation of a specific RF II species. J. Mol. Biol. *72*, 51–63 (1972)

Fidanián, H.M., Ray, D.S.: Replication of bacteriophage M13. VIII. Differential effects of rifampicin and nalidixic acid on the synthesis of the two strands of M13 duplex DNA. J. Mol. Biol. *83*, 63–82 (1974)

Flügel, R.M., Rapp, U., Wells, R.D.: RNA-DNA covalent bonds between RNA primers and DNA products formed by RNA tumor-virus DNA-polymerase. J. Virol. *12*, 1491–1502 (1973)

Fox, R.M., Mendelsohn, J., Barbosa, E., Goulian, M.: RNA in nascent DNA from cultured human lymphocytes. Nature (London) New Biol. *245*, 234–237 (1973)

Francke, B., Ray, D.S.: Cis-limited action of the gene-A-product of bacteriophage φX174, and the essential bacterial site. Proc. Natl. Acad. Sci. *69*, 475–479 (1972)

Fujisawa, H., Hayashi, M.: Gene A product of φX174 is required for site-specific endonucleolytic cleavage during single-stranded DNA synthesis in vivo. J. Virol. *19*, 416–421 (1976)

Gautschi, J.R., Clarkson, J.M.: Discontinuous DNA-repliction in mouse P-185 cells. Eur. J. Biochem. *50*, 403–412 (1975)

Gefter, M.L.: DNA replication. Ann. Rev. Biochem. *44*, 45–78 (1975)

Gefter, M.L., Sherman, L.A.: The role of DNA structure in DNA replication. In: The Organization and Expression of the Eucaryotic Genome, *Bradbury, E.M., Javaherian, K.* (Eds.), London: Academic Press, 1977, pp. 233–248

Geider, K.: Molecular aspects of DNA replication in *Escherichia coli*. Curr. Top. Microbiol. Immunol. *74*, 55–112 (1976)

Geider, K., Beck, E., Schaller, H.: An RNA transcribed from DNA at the origin of phage fd single-strand to replicative form conversion. Proc. Natl. Acad. Sci. *75*, 645–649 (1978)

Geider, K., Kornberg, A.: Conversion of the M13 viral single strand to the double-stranded replicative forms by purified proteins. J. Biol. Chem. *249*, 3999–4005 (1974)

Gellert, M., Mizuchi, K., O'Dea, M.H., Nash, H.A.: DNA gyrase: An enzyme that introduces superhelical turns into DNA. Proc. Natl. Acad. Sci. *73*, 3872–3876 (1976)

Ghysen, A., Pironio, M.: Relationship between the N function of bacteriophage λ and host RNA polymerase. J. Mol. Biol. *65*, 259–272 (1972)

Gilbert, W.: Starting and stopping sequences for the RNA polymerase. In: RNA Polymerase, *Losick, R., Chamberlin, M.* (Eds.), Cold Spring Harbor: Cold Spring Harbor Laboratory, 1976, pp. 193–205

Godson, G.M.: Evolution of φX174. Isolation of four new φX-like phages and comparison with φX174. Virology *58*, 272–289 (1974)

Goebel, W., Schrempf, H.: Possible involvement of DNA polymerase I in excision of RNA from Col E1 DNA *in vivo*. Nature (London) New Biol. *245*, 39–41 (1973)

Gray, C.P., Sommer, R., Polke, C., Beck, E., Schaller, H.: Structure of the origin of DNA replication of bacteriophage fd. Proc. Natl. Acad. Sci. *75*, 50–53 (1978)

Gross, J.D.: DNA replication in bacteria. Curr. Top. Microbiol. Immunol. *57*, 39–74 (1972)

Goulian, M.: Initiation of the replication of single-stranded DNA by *Escherichia coli* DNA polymerase. Cold Spring Harbor Symp. Quant. Biol. *33*, 11–20 (1968)

Harada, F., Sawyer, R.C., Dahlberg, J.E.: Primer ribonucleic acid for initiation of *in vitro* Rous-sarcoma virus deoxyribonucleic acid synthesis—nucleotide sequence and amino-acid acceptor activity. J. Biol. Chem. *250*, 3487–3497 (1975)

Hartman, D., Werner, R.: In vitro transfer of [^{32}P] label from [α-^{32}P] dATP into ribonucleotides. Biochem. Biophys. Res. Commun. *77*, 1445–1451 (1977)

Hartmann, G., Honikel, K.O., Knüsel, F., Nüesch, J.: The specific inhibition of the DNA-directed RNA synthesis by rifamycin. Biochim. Biophys. Acta (Amst.) *145*, 843–844 (1967)

Haseltine, W.A., Kleid, D.G., Panet, A., Rothenberg, E., Baltimore, D.: Ordered transcription of RNA tumor virus genomes. J. Mol. Biol. *106*, 109–132 (1976)

Hausen, P., Stein, H.: Ribonuclease H. An enzyme degrading the RNA moiety of DNA-RNA hybrids. Eur. J. Biochem. *14*, 278–283 (1970)

Hayes, S., Szybalski, W.: Possible RNA primer or DNA replication in coliphage lambda Fed. Proc. *32*, 529Abs (1973)

Hayes, S., Szybalski, W.: Role of oop RNA primer in initiation of coliphage lambda DNA replication. In: DNA Synthesis and its Regulation, *Goulian, M., Hanawalt, P., Fox, C.F.* (Eds.), Menlo Park, Calif.: W.A. Benjamin, Inc. , 1975, pp. 486–512

Heil, A., Zillig, W.: Reconstitution of bacterial DNA-dependent RNA-polymerase from isolated subunits as a tool for the elucidation of the role of the subunits in transcription. FEBS Lett. *11*, 165–168 (1970)

Helinski, D.R.: Plasmid DNA replication. Fed. Proc. *35*, 2026–2030 (1976)

Helinski, D.R., Lovett, M.A., Williams, P.H., Katz, L., Collins, J., Kupersztoch-Portnoy, Y., Sato, S., Leavitt, R.W., Sparks, R., Hershfield, V., Guiney, D.G., Blair, D.G.: Modes of plasmid DNA replication in *Escherichia coli*. In: DNA Synthesis and its Regulation, *Goulian, M., Hanawalt, P., Fox, C.F.* (Eds.), Menlo Park, Calif.: W.A. Benjamin, Inc., 1975, 514–536

Henry, C.M., Ferdinand, F.-J., Knippers, R.: A hybridase from *Escherichia coli*. Biochem. Biophys. Res. Commun. *50*, 603–611 (1973)

Henry, T.J., Knippers, R.: Isolation and function of the gene A initiator of bacteriophage φX174. A highly specific DNA endonuclease. Proc. Natl. Acad. Sci. *71*, 1549–1553 (1974)

Hinkle, D.C., Richardson, C.C.: Bacteriophage T7 deoxyribonucleic acid replication *in vitor*. Requirements for deoxyribonucleic acid synthesis and characterization of the product. J. Biol. Chem. *249*, 2974–2984 (1974)

Hinkle, D.C., Richardson, C.C.: Bacteriophage T7 deoxyribonucleic acid replication *in vitro*. Purification and properties of gene 4 protein of bacteriophage T7. J. Biol. Chem. *250*, 5523–5529 (1975)

Hiraga, S., Saitoh, T.: Initiation of DNA replication in *Escherichia coli*. II. Effect of rifampicin on the resumption of replication of F episome and chromosome upon the returning of *dna* mutants from a nonpermissive to a permissive temperature. Mol. Gen. Genet. *137*, 239–248 (1975)

Hirose, S., Okazaki, R., Tamanoi, F.: Mechanism of DNA chain growth. XI. Structure of RNA-linked DNA fragments of *Escherichia coli.* J. Mol. Biol. *77,* 501–517 (1973)

Hobom, B., Hobom, G.: Rifampicin insensitive replication of dimer lambda dv DNA. Nature (London) New Biol. *244,* 265–267 (1973)

Hohlfeld, R., Vielmetter, W.: Bidirectional growth of the *E. coli* chromosome. Nature (London) New Biol. *242,* 130–132 (1973)

Horiuchi, K., Zinder, N.D.: Origin and direction of synthesis of bacteriophage f1 DNA. Proc. Natl. Acad. Sci. *73,* 2341–2345 (1976)

Hunter, T., Francke, B.: In vitro polyoma DNA synthesis: Involvement of RNA in discontinuous chain growth. J. Mol. Biol. *83,* 123–130 (1974)

Hurwitz, J., Leis, J.P.: RNA-dependent DNA polymerase-activity of RNA tumor viruses. 1. Directing influence of DNA in reaction. J. Virol. *9,* 116–129 (1972)

Ikeda, J.E., Yudelevich, A., Hurwitz, J.: Isolation and characterization of protein coded by gene-A of bacteriophage φX174 DNA. Proc. Natl. Acad. Sci. *73,* 2669–2673 (1976)

Imae, Y., Okazaki, R.: Replication of bacteriophage-T4 DNA *in vitro.* 1. Basic properties of the system. J. Virol. *19,* 435–445 (1976)

Inselburg, J.: Replication of colicin E1 plasmid DNA in minicells from a unique replication initiation site. Proc. Natl. Acad. Sci. *71,* 2256–2259 (1974)

Iwakura, Y., Ishihama, A., Yura, T.: RNA polymerase mutants of .E. coli. II. Streptolydigin resistance and its relation to rifampicin resistance. Mol. Gen. Genet. *121,* 181–196 (1973)

Josse, J., Kaiser, A.D., Kornberg, A.: Enzymatic synthesis of nucleic acid: VIII. Frequencies of nearest neighbor base sequences in deoxyribonucleic acid. J. Biol. Chem. *236,* 864–875 (1961)

Jovin, T.M.: Recognition mechanism of DNA-specific enzymes. Ann. Rev. Biochem. *45,* 889–920 (1976)

Junghans, R.P., Duesberg, P.H., Knight, C.A.: In vitro synthesis of full-length DNA transcripts of Rous sarcoma virus RNA by viral DNA polymerase. Proc. Natl. Acad. Sci. *72,* 4895–4899 (1975)

Karkas, J.D., Chargaff, E.: Template functions in the enzymic formation of polyribonucleotides. III. Apurinic acid as template. Proc. Natl. Acad. Sci. *56,* 1241–1246 (1966)

Kaufmann, G., Anderson, S., Depamphilis, M.L.: RNA primers in Simian virus 40 DNA replication. II. Distribution of 5′ terminal oligoribonucleotides in nascent DNA. J. Mol. Biol. *116,* 549–567 (1977)

Keller, W.: RNA-primed DNA synthesis in vitro. Proc. Natl. Acad. Sci. *69,* 1560–1564 (1972)

Keller, W., Crouch, R.: Degradation of DNA:RNA hybrids by ribonuclease H and DNA polymerases of cellular and viral origin. Proc. Natl. Acad. Sci. *69,* 3360–3364 (1972)

King, A.M.Q.: High molecular weight RNAs from Rous sarcoma virus and Moloney murine leukemia virus contain two subunits. J. Biol. Chem. *251,* 141–149 (1976)

Klein, A., Powling, A.: Initiation of λ DNA replication *in vitro.* Nature (London) New Biol. *239,* 71–73 (1972)

Klein, A., Nüsslein, V., Otto, B., Powling, A.: In vitro studies on *Escherichia coli* DNA replication factors and on initiation of phage λ DNA replication. In: DNA synthesis *in vitro, Wells, R.D., Inman, R.B.* (Eds.), Baltimore-London-Tokyo: University Park Press, 1973, pp. 185–194

Kline, B.C.: Inhibition of plasmid DNA replication by rifampicin in *Salmonella pullorum.* Biochem. Biophys. Res. Commun. *46,* 2019–2025 (1972)

Kline, B.C.: Role of DNA transcription in the initiation of *Escherichia coli* sex factor (F) DNA replication. Biochem. Biophys. Res. Commun. *50,* 280–288 (1973)

Kolodner, R., Richardson, C.C.: Replication of duplex DNA by bacteriophage T7 DNA polymerase and gene 4 protein is accompanied by hydrolysis of nucleoside 5′-triphosphates. Proc. Natl. Acad. Sci. *74,* 1525–1529 (1977)

Konrad, E.B., Lehman, I.R.: A conditional lethal mutant of *Escherichia coli* K12 defective in the 5′-3′ exonuclease associated with DNA polymerase I. Proc. Natl. Acad. Sci. *71,* 2048–2051 (1974)

Konrad, E.B., Lehman, I.R.: Novel mutants of *Escherichia coli* that accumulate very small DNA replicative intermediates. Proc. Natl. Acad. Sci. *72,* 2150–2154 (1975)

Konrad, E.B., Modrich, P., Lehman, I.R.: DNA synthesis in strains of *Escherichia coli* K12 with temperature-sensitive DNA ligase and DNA polymerase I. J. Mol. Biol. *90,* 115–126 (1974)

Kornberg, A.: DNA Synthesis. San Francisco: W.H. Freeman, 1974

Kornberg, A.: RNA priming of DNA synthesis. In: RNA Polymerase, *Losick, R., Chamberlin, M.* (Eds.). Cold Spring Harbor: Cold Spring Harbor Laboratory, 1976, pp. 331–352

Kornberg, A.: Multiple stages in the enzymic replication of DNA. Biochem. Soc. Trans. *5,* 359–374 (1977)

Kozinski, A.W.: Molecular recombination in the ligase negative T4 amber mutant. Cold Spring Harbor Symp. Quant. Biol. *33,* 375–391 (1968)

Kurosawa, Y., Okazaki, R.: Mechanism of DNA chain growth. 13. Evidence for discontinuous replication of both strands of P2 phage DNA. J. Mol. Biol. *94,* 229–241 (1975)

Kurosawa, Y., Ogawa, T., Hirose, S., Okazaki, T., Okazaki, R.: Mechanism of DNA chain growth. 15. RNA-linked nascent DNA pieces in *Escherichia coli* strains assayed with spleen exonuclease. J. Mol. Biol. *96,* 653–664 (1975)

Lark, K.G.: Genetic control over the initiation of the synthesis of the short deoxynucleotide chains. in *E. coli.* Nature (London) New Biol. *240,* 237–240 (1972a)

Lark, K.G.: Evidence for the direct involvement of RNA in the initiation of DNA replication in *Escherichia coli* 15T⁻. J. Mol. Biol. *64,* 47–60 (1972b)

Lehman, I.R.: DNA ligase: Structure, mechanism, and function. Science *186,* 790–797 (1974)

Lehman, I.R., Uyemura, D.G.: DNA polymerase I.: Essential replication enzyme. Coordination of polymerization and 5′–3′ exonuclease is an essential feature for discontinuous DNA replication. Science *193,* 963–969 (1976)

Leis, J.P., Hurwitz, J.: RNA-dependent DNA polymerase activity of RNA tumor viruses. 2. Directing influence of RNA in the reaction. J. Virol. *9,* 130–142 (1972)

Leis, J.P., Berkower, I., Hurwitz, J.: Mechanism of action of ribonuclease H isolated from avian myeloblastosis virus and *Escherichia coli.* Proc. Natl. Acad. Sci. *70,* 466–470 (1973)

Lin, N.S.-C., Pratt, D.: Role of bacteriophage M13 gene 2 in viral DNA replication. J. Mol. Biol. *72,* 37–49 (1972)

Lindahl, T.: New class of enzymes acting on damaged DNA. Nature (London) *259,* 64–66 (1976)

Livingston, D.M., Richardson, C.C.: Deoxyribonucleic acid polymerase III of *Escherichia coli.* Characterization of associated exonuclease activities. J. Biol. Chem. *250,* 470–478 (1975)

Louarn, J., Funderburgh, M., Bird, R.E.: More precise mapping of the replication origin in *Escherichia coli* K12. J. Bacteriol. *120,* 1–5 (1974)

Lovett, M.A., Katz, L., Helinski, D.R.: Unidirectional replication of plasmid Col E1 DNA. Nature (London) *251,* 337–340 (1974)

Magnusson, G., Pigiet, V., Winnacker, E.L., Abrams, R., Reichard, P.: RNA-linked short DNA fragments during polyoma replication. Proc. Natl. Acad. Sci. *70,* 412–415 (1973)

Maitra, U.: Induction of a new RNA polymerase in *Escherichia coli* infected with bacteriophage T3. Biochem. Biophys. Res. Commun. *43,* 443–450 (1971)

Maitra, U., Hurwitz, J.: The role of DNA in RNA synthesis. IX. Nucleoside triphosphate termini in RNA polymerase products. Proc. Natl. Acad. Sci. *54,* 815–822 (1965)

Marsh, R.C., Worcel, A.: A DNA fragment containing the origin of replication of the *Escherichia coli* chromosome. Proc. Natl. Acad. Sci. *74,* 2720–2724 (1977)

Marvin, D.A., Hohn, B.: Filamentous bacterial viruses. Bacteriol. Rev. *33,* 172–209 (1969)

Masker, W.E., Richardson, C.C.: Bacteriophage T7 deoxyribonucleic acid replication *in vitro.* 6. Synthesis of biologically active T7 DNA. J. Mol. Biol. *100,* 557–567 (1976)

diMauro, E., Snyder, L., Marino, P., Lamberti, A., Coppo, A., Tocchini-Valentini, G.P.: Rifampicin sensitivity of the components of DNA-dependent RNA polymerase. Nature (London) *222,* 533–537 (1969)

Maxam, A.M., Gilbert, W.: A new method for sequencing DNA. Proc. Natl. Acad. Sci. *74,* 560–564 (1977)

McConaughy, B.L., Laird, C.D., McCarthy, B.J.: Nucleic acid reassociation in formamide. Biochemistry *8*, 3289–3294 (1969)

McGhee, J.D., von Hippel, P.H.: Formaldehyde as a probe of DNA structure. 3. Equilibrium denaturation of DNA and synthetic polynucleotides. Biochemistry *16*, 3267–3276 (1977)

McFadden, G., Denhardt, D.T.: Mechanism of replication of φX174 single-stranded DNA. IX. Requirement for the *Escherichia coli dna*G protein. J. Virol. *14*, 1070–1075 (1974)

McKenna, W.G., Masters, M.: Biochemical evidence for the bidirectional replication of DNA in *Escherichia coli.* Nature (London) *240*, 536–539 (1972)

McMacken, R., Bouché, J.-P., Rowen, S.L., Weiner, J.H., Ueda, K., Thelander, L., McHenry, C., Kornberg, A.: RNA priming of DNA replication. In: Nucleic Acid-Protein Recognition, *Vogel, H.J.* (Ed.). New York: Academic Press, 1977, pp. 15–29

McMacken, R., Weiner, J.: Multiple stages in the conversion of φX174 DNA to the replicative form. Fed. Proc. *35*, 1407–1407 (1976)

McMacken, R., Ueda, K., Kornberg, A.: Migration of *Escherichia coli dna*B protein on the template DNA-strand as a mechanism in initiating DNA replication. Proc. Natl. Acad. Sci. *74*, 4190–4194 (1977)

Mendelsohn, J., Castagnola, J.M., Goulian, M.: On the mechanism for formation of RNA:DNA complexes from lymphocytes. Biochim. Biophys. Acta *407*, 283–291 (1975)

Messer, W.: Initiation of deoxyribonucleic acid replication in *Escherichia coli* B/r: Chronology of events and transcriptional control of initiation. J. Bacteriol. *112*, 7–12 (1972)

Messer, W., Dankwarth, L., Tippe-Schindler, M., Womack, J., Zahn, G.: Regulation of the initiation of DNA replication in *E. coli:* Alteration of o-RNA and the control of o-RNA synthesis. In: DNA Synthesis and its Regulation. *Goulian, M., Hanawalt, P., Fox, C.F.* (Eds.), Menlo Park, Calif.: W.A. Benjamin, Inc., 1975, pp. 602–617

Messing, J., Staudenbauer, W.O., Hofschneider, P.H.: Inhibition of minicircular DNA replication in *Escherichia coli* 15 by rifampicin. Nature (London) New Biol. *238*, 202–203 (1972)

Miller, H.I., Riggs, A.D., Gill, G.N.: Ribonuclease H (hybrid) in *Escherichia coli.* Identification and characterization. J. Biol. Chem. *248*, 2621–2624 (1973)

Miller, Jr., R.C.: Asymmetric annealing of an RNA linked DNA molecule isolated during the initiation of bacteriophage T7 DNA replication. Biochem. Biophys. Res. Commun. *49*, 1082–1086 (1972)

Mölling, K., Bolognesi, D.P., Bauer, H., Büsen, W., Plassmann, H.W., Hausen, P.: Association of viral reverse transcriptase with an enzyme degrading the RNA moiety of RNA-DNA hybrids. Nature (London) New Biol. *234*, 240–243 (1971)

Morris, C.F., Sinha, N.K., Alberts, B.M.: Reconstitution of bacteriophage T4 DNA replication apparatus from purified components: Rolling circle replication following *de novo* chain initiation on a single-stranded circular DNA template. Proc. Natl. Acad. Sci. *72*, 4800–4804 (1975)

Moses, R.E., Richardson, C.C.: Replication and repair of DNA in cells of *Escherichia coli* treated with toluene. Proc. Natl. Acad. Sci. *67*, 674–681 (1970)

Mosig, G.: A preferred origin and direction of bacteriophage T4 DNA replication. I. A gradient of allele frequencies in crosses between normal and small T4 particles. J. Mol. Biol. *53*, 503–514 (1970)

Nathans, D., Danna, K.J.: Specific origin in SV 40 DNA replication. Nature (London) New Biol. *236*, 200–202 (1972)

Ogawa, T., Hirose, S., Okazaki, T., Okazaki, R.: Mechanism of DNA chain growth. 16. Analyses of RNA-linked DNA pieces in *Escherichia coli* with polynucleotide kinase. J. Mol. Biol. *112*, 121–140 (1977)

Ohasa, S., Tsugita, A.: Purification and characterization of a new ribonucleotide synthesizing enzyme from *Escherichia coli.* J. Mol. Biol. *105*, 545–565 (1976)

Okazaki, T., Okazaki, R.: Mechanism of DNA chain growth. IV. Direction of synthesis of T4 short DNA chains as revealed by exonucleolytic degradation. Proc. Natl. Acad. Sci. *64*, 1242–1248 (1969)

Okazaki, R., Hirose, S., Okazaki, T., Ogawa, T., Kurosawa, Y.: Assay of RNA-linked nascent DNA pieces with polynucleotide kinase. Biochem. Biophys. Res. Commun. *62*, 1018–1024 (1975a)

Okazaki, R., Okazaki, T., Hirose, S., Sugino, A., Ogawa, T., Kurosawa, Y., Shinozaki, K., Tamanoi, F., Seki, T., Machida, Y., Fujiyama, A., Kohara, Y.: Discontinuous replication in procaryotic systems. In: DNA Synthesis and its Regulation. *Goulian, M., Hanawalt, P., Fox, C.F.* (Eds.), Menlo Park, Calif.: W.A. Benjamin, Inc., 1975b, pp. 832–862

Okazaki, R., Sugino, A., Hirose, S., Okazaki, T., Imae, Y., Kainuma-Kuroda, R., Ogawa, I., Arisawa,M., Kurosawa, Y.: The discontinuous replication of DNA. In: DNA synthesis *in vitro, Wells, R.D., Inman, R.B.* (Eds.), Baltimore-London-Tokyo: University Park Press, 1973, pp. 83–106

Olgiati, D.D.G., Pogo, B.G.T., Dales, S.: Evidence for RNA linked to nascent DNA in HeLa-cells. J. Cell Biol. *68*, 557–566 (1976)

Olivera, B.M., Bonhoeffer, F.: Replication of *Escherichia coli* requires DNA polymerase I. Nature (London) *250*, 513–514 (1974)

Pearson, C.K., Davis, P.B., Taylor, A., Amos, N.A.: The involvement of RNA in the initiation of DNA synthesis in mammalian cells: Artifacts arising during caesium-salt density-gradient centrifugation which simulate a covalent attachment of RNA to newly synthesized DNA. Eur. J. Biochem. *62*, 451–459 (1976)

Pesce, A., Casoli, C., Schito, G.C.: Rifampicin-resistant RNA polymerase and NAD transferase activities in coliphage N4 virions. Nature (London) *262*, 412–414 (1976)

Peters, G., Harada, F., Dahlberg, J.E., Panet, A., Haseltine, W.A., Baltimore, D.: Low-molecular-weight RNAs of Moloney murine leukemia-virus—identification of primer for RNA-directed DNA synthesis. J. Virol. *21*, 1031–1041 (1977)

Pigiet, V., Eliasson, R., Reichard, P.: Replication of polyoma DNA in isolated nuclei. III. The nucleotide sequence at the RNA-DNA junction of nascent strands. J. Mol. Biol. *84*, 197–216 (1974)

Pigiet, V., Winnacker, E.L., Eliasson, R., Reichard, P.: Discontinuous elongation of both strands at replication forks in polyoma DNA replication. Nature (London) New Biol. *245*, 203–205 (1973)

Pisetsky, D., Berkower, I., Wickner, R., Hurwitz, J.: Role of ATP in DNA synthesis in *Escherichia coli.* J. Mol. Biol. *71*, 557–571 (1972)

Plevani, P., Chang, L.M.S.: Enzymatic initiation of DNA synthesis by yeast DNA polymerases. Proc. Natl. Acad. Sci. *74*, 1937–1941 (1977)

Probst, H., Gentner, P.R., Hofstätter, T., Jenke, S.: Newly synthesized mammalian cell DNA: Evidence for effects simulating the presence of RNA in the nascent DNA fraction isolated by nitrocellulose column chromatography. Biochim. Biophys. Acta *340*, 361–373 (1974)

Quade, K., Smith, R.E., Nichols, J.L.: Evidence for common nucleotide sequence in the RNA subunits comprising Rous sarcoma virus 70S RNA. Virology *61*, 287–291 (1974)

Ramareddy, G.V., Goulian, S.H., Goulian, M.: DNA synthesis in *in vitro* lysates of *Escherichia coli.* Biochim. Biophys. Acta *402*, 323–342 (1975)

Ray, D.S.: Replication of filamentous bacteriophages. In: Comprehensive Virology, Vol. 7. Reproduction: Bacterial DNA viruses. *Fraenkel-Conrat, H., Wagner, R.R.* (Eds.). New York: Plenum Publishing comp., 1977, pp. 105–178

Ray, D.S., Dueber, J., Suggs, S.: Replication of bacteriophage M13. IX. Requirement of the *Escherichia coli dna*G function for M13 duplex DNA replication. J. Virol. *16*, 348–355 (1975)

Ray, R., Capon, D., Gefter, M.: Distinguishable steps in the enzymatic synthesis of bacteriophage φX174 replicative form *in vitro.* Biochem. Biophys. Res. Commun. *70*, 506–512 (1976)

Reichard, P., Eliasson, R., Söderman, G.: Initiator RNA in discontinuous polyoma DNA synthesis. Proc. Natl. Acad. Sci. *71*, 4901–4905 (1974)

Riva, S., Silvestri, L.: Rifamycins: A general review. Ann. Rev. Microbiol. *26*, 199–224 (1972)

Rowen, S.L., Bouché, J.P.: RNA primer for DNA replication synthesized by *dna*G RNA polymerase. Fed. Proc. *35*, 1418–1418 (1976)

Rowen, L., Kornberg, A.: Primase, the *dna*G protein of *Escherichia coli.* An enzyme which starts DNA chains. J. Biol. Chem. *253*, 758–764 (1978a)

Rowen, L., Kornberg, A.: A ribo-deoxyribonucleotide primer synthesized by primase. J. Biol. Chem. *253,* 770–774 (1978b)

Roychoudhury, R., Kössel, H.: Transcriptional role in DNA replication: Degradation of RNA primer during DNA synthesis. Biochem. Biophys. Res. Commun. *50,* 259–265 (1973)

Sadoff, R.B., Cheevers, W.P.: Evidence for RNA-linked nascent strands in polyoma virus DNA replication. Biochem. Biophys. Res. Commun. *53,* 818–823 (1973)

Sakai, H., Hashimoto, S., Komano, T.: Replication of deoxyribonucleic acid in *Escherichia coli* C mutants temperature sensitive in the initiation of chromosome replication. J. Bacteriol.. *119,* 811–820 (1974)

Sakakibara, Y., Tomizawa, J.-I.: Replication of colicin E1 plasmid DNA in cell extracts. Proc. Natl. Acad. Sci. *71,* 802–806 (1974a)

Sakakibara, Y., Tomizawa, J.-I.: Replication of colicin E1 plasmid DNA in cell extracts: II. Selective synthesis of early replicative intermediates. Proc. Natl. Acad. Sci. *71,* 1403–1407 (1974b)

van de Sande, J.H., Kleppe, K., Khorana, H.G.: Reversal of bacteriophage T4 induced polynucleotide kinase action. Biochemistry *12,* 5050–5055 (1973)

Sato, S., Ariakne, S., Saito, M., Sugimura, T.: RNA bound to nascent DNA in Ehrlich Ascites tumor cells. Biochem. Biophys. Res. Commun. *49,* 827–834 (1972)

Sawyer, R.C., Dahlberg, J.E.: Small RNAs of Rous sarcoma virus: Characterization by two-dimensional polyacrylamide gel electrophoresis and fingerprint analysis. J. Virol. *12,* 1226–1237 (1973)

Sawyer, R.C., Harada, F., Dahlberg, J.E.: Virion-associated RNA primer for Rous sarcoma virus DNA synthesis—isolation from uninfected cells. J. Virol. *13,* 1302–1311 (1974)

Schaller, H., Uhlmann, A., Geider, K.: A DNA fragment from the origin of single-strand to double-strand DNA replication of bacteriophage fd. Proc. Natl. Acad. Sci. *73,* 49–53 (1976)

Schekman, R., Weiner, A., Kornberg, A.: Multienzyme systems of DNA replication. Science *186,* 987–993 (1974)

Schekman, R., Wickner, W., Westergaard, O., Brutlag, D., Geider, K., Bertsch, L.L., Kornberg, A.: Initiation of DNA synthesis of φX174 replicative form requires RNA synthesis resistant to rifampicin. Proc. Natl. Acad. Sci. *69,* 2691–2695 (1972)

Scherer, G., Hobom, G., Kössel, H.: DNA base sequence of the p_0 promoter region of phage λ. Nature (London) *265,* 117–121 (1977)

Scherzinger, E., Klotz, G.: Studies on bacteriophage T7 DNA synthesis *in vitro*: II. Reconstitution of the T7 replication system using purified proteins. Mol. Gen. Genet. *141,* 233–249 (1975)

Scherzinger, E., Lanka, E., Hillenbrand, G.: Role of bacteriophage T7 DNA primase in the initiation of DNA strand synthesis. Nucl. Acid Res. *4,* 4151–4163 (1977b)

Scherzinger, E., Litfin, F.: *In vitro* studies on the role of phage T7 gene 4 product in DNA replication. Mol. Gen. Genet. *135,* 73–86 (1974)

Scherzinger, E., Lanka, E., Morelli, G., Seiffert, D., Yuki, A.: Bacteriophage-T7-induced DNA-priming protein. Eur. J. Biochem. *72,* 543–548 (1977)

Schleif, R.: Isolation and characterization of a streptolydigin resistant RNA polymerase. Nature (London) *223,* 1068–1069 (1969)

Schnös, M., Inman, R.B.: Position of branch points in replicating λ DNA. J. Mol. Biol. *51,* 61–73 (1970)

Schnös, M., Inman, R.B.: The starting point and direction of replication in P2 DNA. J. Mol. Biol. *55,* 31–38 (1971)

Scott, J.F., Eisenberg, S., Bertsch, L.L., Kornberg, A.: A mechanism of duplex DNA replication revealed by enzymatic studies of phage φX174: Catalytic strand separation in advance of replication. Proc. Natl. Acad. Sci. *74,* 193–197 (1977)

Shine, J., Czernilofsky, A.P., Friedrich, R., Bishop, J.M., Goodman, H.M.: Nucleotide sequence at the 5′-terminus of the avian sarcoma virus genome. Proc. Natl. Acad. Sci. *74,* 1473–1477 (1977)

Sinsheimer, R.L.: Bacteriophage φX174 and related viruses. In: Progr. in Nucl. Acid Res. Mol. Biol. *8,* 115–169 (1968)

Silverstein, S., Billen, D.: Transcription: Role in the initiation and replication of DNA synthesis in *Escherichia coli* and φX174. Biochim. Biophys. Acta *247*, 383–390 (1971)

Sobell, H.M., Jain, S.C.: Stereochemistry of actinomycin binding to DNA. II. Detailed molecular model of actinomycin-DNA complex and its implications. J. Mol. Biol. *68*, 21–34 (1972)

Speyer, J.F., Chao, J., Chao, L.: Ribonucleotides covalently linked to deoxyribonucleic acid in T4-bacteriophage. J. Virol. *10*, 902–908 (1972)

Strätling, W., Ferdinand, F.J., Krause, E., Knippers, R.: Bacteriophage T7-DNA replication *in vitro*: an experimental system. Eur. J. Biochem. *38*, 160–169 (1973)

Staudenbauer, W.L.: Replication of colicinogenic factor E1 DNA in plasmolysed *Escherichia coli* cells—coupling of DNA replication and RNA synthesis. Eur. J. Biochem. *58*, 303–313 (1975)

Staudenbauer, W.L., Hofschneider, P.H.: Replication of bacteriophage M13—inhibition of single-strand DNA synthesis by rifampicin. Proc. Natl. Acad. Sci. *69*, 1634–1637 (1972)

Stavrianopoulos, J.G., Karkas, J.D., Chargaff, E.: Mechanism of DNA replication by highly purified DNA polymerase of chicken embryo. Proc. Natl. Acad. Sci. *69*, 2609–2613 (1972)

Sternglanz, R., Wang, H.F., Donegan, J.J.: Evidence that both growing DNA chains at a replication fork are synthesized discontinuously. Biochemistry *15*, 1838–1843 (1976)

Stevens, W.F., Adhya, S., Szybalski, W.: Origin and bidirectional orientation of DNA replication in coliphage lambda. In: The Bacteriophage Lambda, *Hershey, A.D.* (Ed.). Cold Spring Harbor: Cold Spring Harbor Laboratory, 1971, pp. 515–533

Suggs, S.V., Ray, D.S.: Replication of bacteriophage M13. XI. Localization of the origin for M13 single-strand synthesis. J. Mol. Biol. *110*, 147–163 (1977)

Sugino, A., Okazaki, R.: Mechanism of DNA chain growth. 7. Direction and rate of growth of T4 nascent short DNA chains. J. Mol. Biol. *64*, 61–85 (1972)

Sugino, A., Okazaki, R.: RNA-linked DNA fragments *in vitro*. Proc. Natl. Acad. Sci. *70*, 88–92 (1973)

Sugino, A., Hirose, S., Okazaki, R.: RNA-linked nascent DNA fragments in *Escherichia coli*. Proc. Natl. Acad. Sci. *69*, 1863–1867 (1972)

Sugino, Y., Tomizawa, J., Kakefuda, T.: Location of non-DNA components of closed circular colicin E1 plasmid DNA. Nature (London) *253*, 652–654 (1975)

Tabak, H.F., Griffith, J., Geider, K., Schaller, H., Kornberg, A.: Initiation of deoxyribonucleic acid synthesis. VII. A unique location of the gap in the M13 replicative duplex synthesized *in vitro*. J. Biol. Chem. *249*, 3049–3054 (1974)

Tamblyn, T.M., Wells, R.D.: Comparative ability of RNA and DNA to prime DNA synthesis *in vitro*—role of sequence, sugar, and structure of template-primer. Biochemistry *14*, 1412–1425 (1975)

Taylor, J.H.: Units of DNA replication in chromosomes of eucaryotes. Int. Rev. Cytol. *37*, 1–20 (1974)

Taylor, J.H., Wu, M., Erickson, L.C., Kurek, M.P.: Replication of DNA in mammalian chromosomes. 3. Size and RNA content of Okazaki fragments. Chromosoma *53*, 175–189 (1975)

Taylor, J.M.: Analysis of the role of transfer-RNA species as primers for transcription into DNA of RNA tumor virus genomes. Biochim. Biophys. Acta *473*, 57–71 (1977)

Taylor, J.M., Cordell-Stewart, B., Rohde, W., Goodman, H.M., Bishop, J.M.: Reassociation of 4S and 5S RNAs with the genome of avian sarcoma virus. Virology *65*, 248–259 (1975)

Taylor, J.M., Illmensee, R.: Site on RNA of an avian sarcoma virus at which primer is bound. J. Virol. *16*, 553–558 (1975)

Temin, H.M.: The cellular and molecular biology of RNA tumor viruses; especially avian leukosis sarcoma viruses, and their relatives. Adv. Cancer Res. *19*, 47–104 (1974)

Temin, H.M., Mizutani, S.: RNA dependent DNA polymerase in virions of Rous sarcoma virus. Nature (London) *226*, 1211–1213 (1970)

Tocchini-Valentini, G.P., Marino, P., Colvill, A.J.: Mutant of *E. coli* containing an altered DNA-dependent RNA polymerase. Nature (London) *220*, 275–276 (1968)

Tomizawa, J., Ohmori, H., Bird, R.E.: Origin of replication of colicin E1 plasmid DNA. Proc. Natl. Acad. Sci. *74*, 1865–1869 (1977)

Tomizawa, J.-I., Sakakibara, Y., Kakefuda, T.: Replication of Colicin E1 plasmid DNA in cell extracts. Origin and direction of replication. Proc. Natl. Acad. Sci. *71*, 2260–2264 (1974)

Tomizawa, J.-I., Sakakibara, Y., Kakefuda, T.: Replication of Colicin E1 plasmid DNA added to cell extracts. Proc. Natl. Acad. Sci. *72*, 1050–1054 (1975)

Towle, H.C., Jolly, J.F., Boezi, J.A.: Purification and characterization of bacteriophage gh-1-induced deoxyribonucleic acid-dependent ribonucleic acid polymerase from *Pseudomonas putida.* J. Biol. Chem. *250*, 1723–1733 (1975)

Truffaut, N., Manheimer, I.: Development of bacteriophage Ha$_2$, a φX174 derivative, in *Escherichia coli* strains carrying a thermosensitive mutation in the *dna*G gene. Biochimie, *57*, 905–915 (1975)

Tseng, B.Y., Goulian, M.: Evidence for covalent association of RNA with nascent DNA in human lymphocytes. J. Mol.. Biol. *99*, 339–346 (1975)

Tseng, B.Y., Goulian, M.: Initiator RNA of discontinuous DNA synthesis in human lymphocytes. Cell *12*, 483–489 (1977)

Tye, B.-K., Lehman, I.R.: Excision repair of uracil incorporated in DNA as a result of a defect in dUTPase. J. Mol. Biol. *117*, 293–306 (1977)

Tye, B.-K., Nyman, P.-O., Lehman, I.R., Hochhauser, S., Weiss, B.: Transient accumulation of Okazaki fragments as a result of uracil incorporation into nascent DNA. Proc. Natl. Acad. Sci. *74*, 154–157 (1977)

Uyemura, D., Eichler, D.C., Lehman, I.R.: Biochemical characterization of mutant forms of DNA polymerase I from *Escherichia coli.* II. The polAex1 mutation. J. Biol. Chem. *251*, 4085–4089 (1976)

Verma, I.M., Meuth, N.L., Bromfeld, E., Manly, K.F., Baltimore, D.: Covalently linked RNA-DNA molecule as initial product of RNA tumor-virus DNA polymerase. Nature (London) New Biol. *233*, 131–134 (1971)

Vicuna, R., Hurwitz, J., Wallace, S., Girard, M.: Selective inhibition of *in vitro* DNA synthesis dependent on φX174 compared with fd DNA. I. Protein requirements for selective inhibition. J. Biol. Chem. *252*, 2524–2533 (1977a)

Vicuna, R., Ikeda, J.E., Hurwitz, J.: Selective inhibition of φX174 RF II compared with fd RF II DNA synthesis *in vitro.* II. Resolution of discrimination reaction into multiple steps. J. Biol. Chem. *252*, 2534–2544 (1977b)

Vosberg, H.-P., Hoffmann-Berling, H.: DNA synthesis in nucleotide-permeable *Escherichia coli* cells. I. Preparation and properties of ether-treated cells. J. Mol. Biol. *58*, 739–753 (1971)

Waqar, M.A., Huberman, J.A.: Evidence for attachment of RNA to pulse-labeled DNA in slime mold *Physarum polycephalum.* Biochem. Biophys. Res. Commun. *51*, 174–180 (1973)

Waqar, M.A., Huberman, J.A.: Covalent linkage between RNA and nascent DNA in slime mold *Physarum polycephalum.* Biochim. Biophys. Acta *383*, 410–420 (1975a)

Waqar, M.A., Huberman, J.A.: Covalent attachment of RNA to nascent DNA in mammalian cells. Cell *6*, 551–557 (1975b)

Waqar, M.A., Minkoff, R., Tsai, A., Huberman, J.A.: Evidence for RNA linked to nascent DNA in eucaryotic organisms. In: DNA Synthesis and its Regulation, *Goulian, M., Hanawalt, P., Fox, C.F.* (Eds.), Menlo Park, Calif.: W.A. Benjamin, Inc., 1975, pp. 334–356

Waters, L.C.: Transfer RNAs associated with the 70S RNA of AKR murine leukemia virus. Biochem. Biophys. Res. Commun. *65*, 1130–1136 (1975)

Waters, L.C., Mullin, B.C.: Comparison of the tRNAs associated with the 70S RNA of various RNA tumor viruses. Fed. Proc. *35*, 1736–1736 (1976)

Waters, L.C., Mullin, B.C.: Transfer RNA in RNA tumor viruses. In Progr. Nucl. Acid Res. Mol. Biol. *20*, 131–160 (1977)

Wechsler, J.A., Gross, J.D.: Escherichia coli mutants temperature-sensitive for DNA synthesis. Mol. Gen. Genet. *113*, 273–284 (1971)

Wehrli, W., Staehelin, M.: Actions of the rifamycins. Bacteriol. Rev. *35*, 290–309 (1971)

Weiner, J.H., McMacken, R., Kornberg, A.: Isolation of an intermediate which precedes *dna*G RNA polymerase participation in enzymatic replication of bacteriophage φX174 DNA. Proc. Natl. Acad. Sci. *73*, 752–756 (1976)

Wells, R.D., Flügel, R.M., Larson, J.E., Schendel, P.F., Sweet, R.W.: Comparison of some reactions catalyzed by deoxyribonucleic acid polymerase from avian myeloblastosis virus, *Escherichia coli*, and *Micrococcus luteus.* Biochemistry *11*, 621–629 (1972)

Westergaard, O., Brutlag, D., Kornberg, A.: Initiation of deoxyribonucleic acid synthesis. IV. Incorporation of the ribonucleic acid primer into the phage replicative form. J. Biol. Chem. *248*, 1361–1364 (1973)

Wickner, R.B.: DNA Replication. New York: Marcel Dekker, 1974

Wickner, R.B., Wright, M., Wickner, S., Hurwitz, J.: Conversion of φX174 and fd single-stranded DNA to replicative forms in extracts of *Escherichia coli.* Proc. Natl. Acad. Sci. *69*, 3233–3237 (1972)

Wickner, S.: DNA or RNA priming of bacteriophage G4 DNA synthesis by *Escherichia coli dna*G protein. Proc. Natl. Acad. Sci. *74*, 2815–2819 (1977)

Wickner, S., Hurwitz, J.: Conversion of φX174 viral DNA to double-stranded form by purified *Escherichia coli* proteins. Proc. Natl. Acad. Sci. *71*, 4120–4124 (1974)

Wickner, S., Hurwitz, J.: Interaction of *Escherichia coli dna*B and *dna*C(D) gene products *in vitro.* Proc. Natl. Acad. Sci. *72*, 921–925 (1975a)

Wickner, S., Hurwitz, J.: In vitro synthesis of DNA. In: DNA Synthesis and its Regulation, *Goulian, M., Hanawalt, P., Fox, C.F.* (Eds.), Menlo Park, Calif.: W.A. Benjamin, Inc., 1975b, pp. 227–238

Wickner, S., Hurwitz, J.: Association of φX174 DNA-dependent ATPase activity with an *Escherichia coli* protein, replication factor Y, required for *in vitro* synthesis of φX174 DNA. Proc. Natl. Acad. Sci. *72*, 3342–3346 (1975c)

Wickner, S., Wright, M., Hurwitz, J.: Studies on *in vitro* DNA synthesis: Purification of the *dna*G gene product from *Escherichia coli.* Proc. Natl. Acad. Sci. *70*, 1613–1618 (1973)

Wickner, S., Wright, M., Hurwitz, J.: Association of DNA-dependent and -independent ribonucleoside triphosphatase activities with *dna*B gene product of *Escherichia coli.* Proc. Natl. Acad. Sci. *71*, 783–787 (1974)

Wickner, W., Kornberg, A.: DNA polymerase III star requires ATP to to start synthesis on a primed DNA. Proc. Natl. Acad. Sci. *70*, 3679–3683 (1973)

Wickner, W., Kornberg, A.: A novel form of RNA polymerase from *Escherichia coli* Proc. Natl. Acad. Sci. *71*, 4425–4428 (1974)

Wickner, W., Brutlag, D., Schekman, R., Kornberg, A.: RNA synthesis initiates *in vitro* conversion of M13 DNA to its replicative form. Proc. Natl. Acad. Sci. *69*, 965–969 (1972)

Williams, P.H., Boyer, H.W., Helinski, D.R.: Size and base composition of RNA in supercoiled plasmid DNA. Proc. Natl. Acad. Sci. *70*, 3744–3748 (1973)

Winnacker, E.L.: Adenovirus type 2 DNA replication. I. Evidence for discontinuous synthesis. J. Virol. *15*, 744–758 (1975)

Wyers, F., Sentenac, A., Fromageot, P.: Role of DNA-RNA hybrids in eucaryotes. Ribonuclease H in yeast. Eur. J. Biochem. *35*, 270–281 (1973)

Zechel, K., Bouché, J.-P., Kornberg, A.: Replication of phage G4: A novel and simple system for the initiation of deoxyribonucleic acid synthesis. J. Biol. Chem. *250*, 4684–4689 (1975)

Zyskind, J.W., Deen, L.T., Smith, D.W.: Temporal sequence of events during the initiation process in *Escherichia coli* deoxyribonucleic acid replication: Roles of the *dna*A and *dna*C gene products and ribonucleic acid polymerase. J. Bacteriol. *129*, 1466–1475 (1977)

Zyskind, J.W., Smith, D.W.: Novel *Escherichia coli* mutant: Direct involvement of the *dna*B252 gene product in the synthesis of an origin-ribonucleic acid species during initiation of a round of deoxyribonucleic acid replication. J. Bacteriol.. *129*, 1476–1486 (1977)

Speculations on the Role of Major Transplantation Antigens in Cell-Mediated Immunity Against Intracellular Parasites

ROLF M. ZINKERNAGEL [1]

Abbreviations: Ag-B = Major rat histocompatibility; CMI = Cell-mediated immunity; DTH = Delayed-type hypersensitivity; H = Major histocompatibility; H-2 = Major murine histocompatibility; HL-A = Major human histocompatibility; Ir = Immune response; P = Parental; T = Thymus derived.

I. Introduction

In three major fields of immunology — "academic" immunology, immunology of infectious diseases, and transplantation immunology — exciting features in

[1] The Department of Cellular and Developmental Immunology, Research Institute of Scripps Clinic, La Jolla, California 92037, USA

common have been discovered over the past few years. As early as the 1950's, possible links between the transplantation reaction and cell-mediated immunity to cell-bound antigens or intracellular parasites were postulated. For instance, Mitchison speculated that cellular immune recognition of antigen in delayed-type hypersensitivity (DTH) reactions probably occurred only when the antigen involved (e.g., tuberculin or chemical allergens) were present on the cell-membrane, thus resembling a foreign transplantation antigen (*Mitchison,* 1954). *Lawrence,* stimulated by *Thomas'* surveillance hypothesis (*Thomas,* 1959), proposed that immune lymphocytes recognize not only foreign antigens on cell surfaces, since cell-mediated immunity originally evolved against intracellular parasites, but also recognized a self-component (*Lawrence,* 1959, 1974). This self-plus-X hypothesis was a very lucid speculation that foresaw many of the principles discovered since. All these discoveries now lead to the conclusion that thymus-derived lymphocytes (T-cells) seem to express a double specificity for foreign antigenic determinants and for cell-surface self determinants (For reviews see *Paul* and *Benacerraf,* 1977; *Munro* and *Bright,* 1976; *Doherty* et al., 1976; *Shearer* et al., 1976; *Forman,* 1976; *Zinkernagel* and *Doherty,* 1977a; *Koszinowski* et al., 1977a; *Schrader* et al., 1977; *Katz,* 1977; *Langman,* 1978).

The facts that substantiate early speculations about the involvement of self cell-surface markers in cell-mediated immunity (CMI) are as follows: *McDevitt* described mice and *Benacerraf* guinea pigs in which immune responses to certain antigens were regulated by genes coding within the major histocompatibility (H) gene complex (*Benacerraf* and *McDevitt,* 1972). Then *Lilly* et al. showed that, in some mouse strains, the susceptibility to induction of viral leukemogenesis segregated with the murine H gene complex (*H-2*) (*Lilly* et al., 1964; *Lilly,* 1968; *Lilly* and *Pincus,* 1973). Subsequently, similar evidence linked disease susceptibility with certain human H (*H-2*) types in humans. (In reviews *Dausset,* 1972; *Amos* et al., 1973; *Morris,* 1974; *Zinkernagel* and *Doherty,* 1976a; *Vladutiu* and *Rose,* 1972.) About at the same time *Bryere* and *Williams* (1964) and *Svet-Moldavsky* and collaborators (*Svet-Moldavsky* et al., 1968) reported that skin from inbred mice that were congenitally infected with oncogenic virus was rejected by syngeneic nonvirus carrier mice, just as allogeneic skin is usually rejected. This phenomenon of heterogenization was also observed by *Lindenmann* and coworkers, who found that virus-infected tumor cells, but not uninfected cells, immunized hosts against subsequent tumor growth (*Lindenmann* and *Klein,* 1967). After the discovery that T-B cells collaborate in the immune response, (*Claman* et al., 1966; *Miller* and *Mitchell,* 1968; *Mitchell* and *Miller,* 1968; *Mitchison,* 1971a, b) *Katz* et al. (1971) found that the need for carrier-specific T cells could be fulfilled by unprimed allogeneic T cells. This "allogeneic effect" preempted some of the subsequent findings by *Kindred* and *Shreffler* (1972) and *Katz* and collaborators that helper T cells had to be *H-2* compatible with B cells to cooperate efficiently (*Katz* et al., 1973; *Katz* and *Benacerraf,* 1975; *Katz,* 1977). These results explained the finding of *Miller* et al. some 4 years earlier that allogeneic T and B cells cooperated poorly (*Miller* and *Mitchell,* 1968; *Mitchell* and *Miller,* 1968). Independently, *Rosenthal* and *Shevach* reported that antigen-specific T cell proliferation was not only antigen-specific but also depended upon responding T cells and stimulating antigen-pulsed macrophages being histocompatible (*Rosenthal* and *Shevach*, 1973; *Shevach* and *Rosenthal*, 1973). Both examples of T cell interactions in mice were subsequently shown to be specific for *H-2I* region coded structures (or the guinea pig equivalent thereof). Since both these T cell-effector functions were under *Ir*-gene control, these examples appeared to offer possible explanations for the mode of action of *Ir* genes as regulators of T cell-B cell or T cell-B cell-macrophage interactions (For reviews see *Katz* and *Benacerraf*, 1975; *Munro* and *Bright*, 1976; *Zinkernagel* and *Doherty*, 1977 b).

The latest decisive finding in conceptualizing the actual role and function of the major histocompatibility gene complex was the *H-2* restriction of the cytotoxic activity of virus-specific T cells (*Zinkernagel* and *Doherty*, 1974 a, b). This *H-2* restriction also applies to trinitrophenyl (TNP) and other hapten or minor alloantigen-specific cytotoxic T cells (*Shearer*, 1974; *Bevan*, 1975 a; *Gordon* et al., 1975). All these examples that established the crucial role of the major histocompatibility gene complex in "syngeneic" reactions, beside its well-established role in allogeneic confrontations (reviewed in *Cerottini* and *Brunner*, 1974), have been reviewed extensively (*Katz* and *Benacerraf*, 1974, *Doherty* et al., 1976 b; *Shearer* et al., 1976; *Forman*, 1976; *Zinkernagel* and *Doherty*, 1977 a, *Langman*, 1978).

The recent preoccupation of immunologists with *H-2* and *H-2* restriction reflects the feeling that essential information on the elusive T cell receptor is hidden in this phenomenon. In mice, all T cell functions that have been tested are *H-2* restricted. Thus, the histocompatibility restriction over immune reactions appears to be a general phenomenon in mice and probably is universal in higher vertebrates. For example, and not surprisingly, virus-specific cytotoxicity in rats is strain-restricted and probably restricted by the rat major histocompatibility gene complex (Ag-B) (*Zinkernagel* et al., 1977 a; *Marshak* et al., 1977). Strong evidence exists that the H-restriction is also operative in chickens as best documented by the elegant experiments of the *Toivanens* (*Toivanen* et al., 1974) who showed that germinal centers in bursectomized chickens could be reconstituted only by transplanting B locus-compatible bursa cells. *Wainberg* and coworkers (*Wainberg* et al., 1974) have also suggested that B locus restriction may be valid for virus-specific cytotoxic T cells in chickens. In humans, *van Rood* and coworkers (*Goulmy* et al., 1976) uncovered the first indication that the *HL-A* restriction may govern cytotoxicity against cells bearing the male H-Y antigen.

This paper is not a comprehensive review, but aims rather at discussing the open questions of T cell recognition and the possible functions of antigens coded by the major histocompatibility gene complex from biologic, teleologic and evolutionary points of view.

II. *H-2* Restriction of Antigen-Specific T Cells

A. Experimental Models and Cellular Parameters

1. Cytotoxic T Cells. First, *Lundstedt* (1969) and subsequently *Oldstone* and *Dixon* (1970) observed that virus-immune lymphocytes specifically inhibited the growth of or destroyed virus-infected syngeneic target cells by direct contact. This observation was analyzed in more detail by *Marker* and *Volker* (1973), *Cole* and coworkers, *Blanden* and *Gardner* and many others (*Cole* and *Nathanson*, 1975; *Blanden*, 1974; *Doherty cand Zinkernagel*, 1974). The characteristics of these highly active virus-immune cytotoxic lymphocytes can be summarized as follows: They are thymus-derived lymphocytes whose cytotoxic function does not involve antibodies. These T cells are virus-specific; they kill by direct contact only in a one hit kinetic pattern. They are demonstrable about 3–5 days after the initiation of infection, peak in quantity at 6–9 days and disappear rapidly thereafter (summarized in *Zinkernagel* and *Doherty*, 1977a). In all these respects, virus-immune cytotoxic T cells have the same characteristics as cytotoxic T cells directed against foreign major transplantation antigens. (For a review see *Cerottini* and *Brunner*, 1974; *Zinkernagel* and *Doherty*, 1977b; *Zinkernagel*, 1977.)

These striking similarities became convincing evidence of the relationship between the transplantation reaction and CMI to intracellular parasites when it was found that virus-specific cytotoxic T cells lysed syngeneic virus-infected target cells but were 30–300 times less efficient in destroying allogeneic target cells infected with the same virus (*Zinkernagel* and *Doherty*, 1974a, b, *Doherty* and *Zinkernagel*, 1975a). Analysis of this restricted virus-specific killing of infected target cells revealed that the restriction mapped to the major murine histocompatibility gene complex (*H-2*), and more precisely to the *K* and *D* but not to the *I* region of *H-2* (*Blanden* et al., 1975a; *Koszinowski* et al., 1977a). Thus alloreactive and syngeneic virus-specific cytotoxic T cells reacted specifically and probably with the same kind of cell surface structures.

Exactly these same *K* and *D* regions of *H-2* were also found to restrict; first, cytotoxic T cells that were generated *in vitro* against trinitrophenylated syngeneic spleen cells (*Shearer* et al., 1975) and, second, cytotoxic T cells that were primed *in vivo* and restimulated *in vitro* against minor histocompatibility antigens (*Bevan*, 1975b; *Gordon* et al., 1975). The basic findings, the experimental conditions, and the characterization of the genetic requirements for cytotoxic interactions, as well as the cellular parameters and their kinetics of generation and activities have been reviewed extensively. Therefore, only selected experiments that appear to hint at some properties of the still-elusive T cell-receptor(s) and mechanisms of T cell recognition are reviewed in the following section.

2. Noncytolytic T Cells. Similar phenomena of T cells that are doubly specific for an antigen plus a self-marker on the target cell were discovered in circumstances marked by notable dissimilarities. In 1972, *Kindred* and *Shreffler* reported that *H-2* -incompatible T cells could not serve as reconstituting T helper cells in nude mice (*Kindred* and *Shreffler*, 1972). This observation was analyzed in great detail by *Katz*, *Benacerraf* and collaborators, *in vivo* and *in vitro* using conventional hapten-carrier antigens (*Katz* et al., 1973a, b; *Katz* et al., 1975). These results can be summarized as follows: T cells collaborate with B cells much more efficiently if the donor of the primed helper T cells and the B cells share the *I* region of *H-2*. Independently, *Rosenthal* and *Shevach*, who assayed guinea pig cells, demonstrated that antigen-specific proliferation of T cells was triggered much more efficiently by antigen-pulsed macrophages of histocompatible donors than of incompatible ones (*Shevach* and *Rosenthal*, 1973; *Rosenthal* and *Shevach*, 1973). *Schwartz* and *Paul* (1976) obtained similar results in the murine system. The adoptive transfer of delayed-type hypersensitivity (DTH) to conventional antigens such as fowl-γ-globulin is *H-2I* restricted in mice (*Miller* et al., 1975) and DTH to chemically reactive antigens or viruses also maps to *K* and *D* (*Miller* et al., 1976). *Erb* and *Feldman* (1975) studied T cell-macrophage interaction in triggering T helper cells. These investigators found antigen-specific macrophage factors that were *H-2I* restricted in their capacity to sensitize T helper-cell activity.

3. Immune Responsiveness. The *H-2* gene complex, besides restricting T cell activities also harbors regulatory genes (immune response genes, *Ir*) as described first by *Benacerraf* and *McDevitt* (1972). Although originally it was found that *Ir* genes influenced only levels of immunoglobulin production against weak antigens, more recently it has become obvious that similar phenomena can be shown for the levels of generation of syngeneic cytotoxic T cells *Schmitt-Verhulst* and *Shearer*, 1975, 1976; *Simpson* and *Gordon*, 1977; *von Boehmer*, 1977). In some instances, two genes seem to be involved in determining responsiveness, one mapping to *I-A* the other probably to the right of *I-C* (*Dorf* and *Benacerraf*, 1975; *Debre* et al., 1976; *Munro* and *Taussig*, 1975; *Shearer* et al., 1976; *Simpson* and *Gordon*, 1977; *von Boehmer*, 1977). The relationship between *H-2* restriction and the two complementing *Ir* genes in *H-2* is unclear as yet. However, future interpretation offer some insight into how T cell recognition occurs, independent of whether these genes code for the T cell receptor(s) (*Benacerraf* and *McDevitt*, 1972), whether such genes define "alterable" self-markers (*Paul* and *Benacerraf*, 1977), whether they code for cell-interaction structures, (*Katz* and *Benacerraf*, 1975) or just for cell-surface markers that have only indirect regulatory function themselves (*Zinkernagel*, 1977).

B. Models of T Cell Recognition

Several models for T cell recognition have been proposed; they fall into one of two groups. First, the dual recognition models and, second, the single recognition models (*Katz* et al., 1973b; *Zinkernagel* and *Doherty*, 1974b; *Doherty* and *Zinkernagel*, 1975b; *Schrader* et al., 1975; *Bevan*, 1975b; *Shearer* et al., 1975; *Blanden* et al., 1976; *Zinkernagel* and *Doherty*, 1977b).

Although the distinction between the two models is in many ways artificial, it has allowed great diversity in experimental designs, approaches and questions. The dual recognition model states that T cells have two receptors, one for the antigen X and one for a self-marker, and these receptors see two distinct antigenic entities.

The single recognition model postulates that T cells have a single receptor that recognizes some sort of a neoantigen formed either by the complex of self-markers with an antigen, by antigen-specific modification of the self-marker or by host-specific modification of the antigen. The single antigen (altered self) recognition model does not permit one to distinguish between these possibilities and a situation in which self-recognition and antigen X recognition are either linked or are independently clonally expressed (*Doherty* et al., 1976a; *Zinker-*

nagel and *Doherty*, 1975). Since both of these latter propositions (linkage or independent clonal expression) are dual recognition models in molecular terms, but appear phenotypically and functionally as one receptor unit, substantial confusion exists in discussing these models. The term *altered self* was originally conceived as a functional concept, but we now prefer to use the term *neoantigen* or new antigenic determinant to define the *single antigenic entity*; *self* or *self-marker* are the antigenic determinants involved in self-recognition and *antigen X* is the conventional type of antigenic determinant. The latter two form the two separate antigenic specificities recognized by a dual receptor.

The various models have been discussed, reviewed and analyzed extensively. Unfortunately, to date there is no clear cut evidence that could exclude or prove one or the other or even a third new model.

III. Experiments to Analyze Mechanisms of T Cells Recognition

1. F_1 Experiments. F_1 heterozygote and *H-2* recombinant mice infected with virus, sensitized against TNP or against minor alloantigens, generate cytotoxic T cells that are specific for one only of the *K* or *D* markers plus antigen. This has been shown in two ways. First F_1 immune cytotoxic T cells proliferate preferentially in the parental one (P_1) recipients that express the foreign antigen to increase cytotoxicity against antigen expressing targets of that same P_1 haplotype (*Zinkernagel* and *Doherty,* 1974b, 1975; *Bevan,* 1976). Second, F_1 T cells that lyse infected P_1 target cells can be inhibited competitively only by excess added antigen expressing targets of the same P_1 haplotype but not by targets of the other parent, P_2 (*Zinkernagel* and *Doherty,* 1975; *Shearer* et al., 1975; *Bevan,* 1976).

These results were initially interpreted to support the single neoantigen-single T cell receptor idea. However, although allelic and genic exclusion of self-recognition was thought to be unlikely (*Zinkernagel* and *Doherty,* 1975) this may in fact be a viable alternative. Thus, these F_1 experiments may be interpreted to reflect that T cells express two receptors that are both clonally expressed; the specificity of self-recognition is regulated by allelic and or genic exclusion mechanisms (*Langman,* 1978; *Zinkernagel* and *Doherty,* 1977a, b; *Zinkernagel,* 1976a).

2. The Chimera Experiments. Several research groups have shown that T cells, differentiated either in chimeras formed by zygote-fusion or in parent (P) →F_1 irradiation chimeras, could cooperate with B cells or could lyse haptenated or virus-infected target cells of the tolerated haplotype from the other parent (*Bechtol* et al., 1974; *Sprent* et al., 1975; *Pfizenmayer* et al., 1976; *Zinkernagel,* 1976a; *von Boehmer* and *Haas,* 1976). This may indicate that these T cells tolerate the other parental haplotype but not the neoantigen formed between it and the haptenic or viral antigens; therefore, the modified alloantigen can be recognized or lysed. However, one can also postulate that such T cells have learned to recognize the tolerated alloantigen as self and handle it as a self-marker as originally proposed by *Katz* and *Benacerraf* (1976) (*Katz,* 1977; *Zinkernagel,* 1976a).

More recent experiments demonstrated that $(H\text{-}2^a \times H\text{-}2^b)F_1 \to P$ $(H\text{-}2^a)$ irradiation bone marrow chimeras generated cytotoxic activity in association with $H\text{-}2^a$ only (*Bevan*, 1977c; *Zinkernagel* et al., 1978a, b, c). In contrast, when such chimeras were formed by using adult spleen cells $(H\text{-}2^a \times H\text{-}2^b)F_1 \to P$ $(H\text{-}2^a)$ then cytotoxic T cells were generated for both $H\text{-}2$ haplotypes. These experiments suggested that the host determined which $H\text{-}2$ specificity lymphopoietic stem-cells could learn to recognize; however, immunocompetent T cells do not change the specificity for self-$H\text{-}2$ under similar conditions. The crucial role of the thymus in dictating the specificity for $H\text{-}2$ of T cells was documented in the following experiment. Adult thymectomized, lethally irradiated and reconstituted mice of $(H\text{-}2^a \times H\text{-}2^b)F_1$ type were transplanted with irradiated thymus lobes of $H\text{-}2^a$ origin. Three months later such thymus chimeras were able to generate significant cytotoxicity against infected target cells of $H\text{-}2^a$ type only. Thus, the radiation-resistant portion of the thymus, i.e., probably the thymic epithelial cells, select the specificity of T cells for $H\text{-}2$ structures expressed on these thymus cells.

These results can be explained along the lines of a single receptor model for a neoantigenic determinant (NAD). Accordingly, precursor T cells may gain their maturity by differentiating receptors for foreign antigens (NAD) in the thymus as proposed by *Jerne* (1971). Thus, in a mouse of $H\text{-}2^a$ type, T cells with specificity for *a* will proliferate during ontogeny in the thymus upon contact with self $H\text{-}2^a$ structures; some of these proliferating T lymphocytes will mutate to express a receptor not any longer specific for *a* but for "slightly different from *a*". Through a "filter mechanism", T cells with specificity for *a* will not be able to leave the thymus and would be pushed to proliferate further, or alternatively are inactivated. Only T cells having a mutated recognition specificity for "different from *a*" will be released from the thymus. This could explain the preference of T cells of $(H\text{-}2^a \times H\text{-}2^b)F_1 \to A$ chimeras for "altered *a*". Since these chimeric T cells cannot react to "altered *b*" we must introduce the rule that the specificity spectrum anti-X that can be generated in a thymus *a* cannot overlap with that generated in a thymus *b*. When applied to a model of single receptor specificity for neoantigenic determinant this rule is unpredicted and seems unlikely but cannot be formally excluded as yet. In the light of the data presented here, which requires the T cell's receptor quality for anti-self-H-2 structures to be specifically learned, i.e., selected for and expressed independently of the receptor quality for anti-X, the dual recognition model provides a simple interpretation. To argue that one receptor has two independent recognition sites is tantamount to accepting two independent recognition sites, i.e., dual recognition. The distinction between models with one and two receptors may, therefore, be reduced to deciding whether signals generated by antigen binding to the anti-self-H-2 structure and anti-X sites are transmitted intracellularly via one or two molecules.

3. Experiments with Mutant Mice. Apparently, there is an inverse relationship between skin graft rejection and the self-recognition capacity of mutant and wildtype mice when tested reciprocally. $H\text{-}2$ mutant mice were selected by skin graft testing y *Bailey, Egorov* and *Kohn* (Reviewed in *Klein,* 1975). The mutant mice's TNP or virus-immune T cells were tested for the capacity to lyse

TNP modified or with virus-infected target cells of the wildtype (*Forman* and *Klein,* 1977; *Zinkernagel,* 1976b; *Doherty* et al., 1976a; *Blanden* et al., 1976). Against TNP-sensitized targets, killer T cells from wildtype mice cross reacted widely with TNP-mutant cells and vice versa. In contrast, virus-specific cytotoxic T cells from some mutnt mice (e.g., the Hzl strain) failed to cross react to various extents. These results can be explained by either theory of T cell recognition (*Zinkernagel* and *Doherty,* 1976b). However, the fact that the cross-reactivity patterns of wildtype and mutant mice are similar for various viruses seems to be most compatible with the dual recognition model (*Zinkernagel* and *Klein,* 1977; *Zinkernagel* and *Doherty,* 1977a). The complete cross reaction of mutant and wildtype *H-2*-restricted TNP-specific cytotoxicity contrasts sharply with the results obtained with virus-specific cytotoxic T cells suggesting that the recognition mechanisms may be different in the two experimental models (*Zinkernagel* and *Klein,* 1977).

4. Similarity of Idiotype of T and B Cell Receptors for Antigens. Many findings indicate that the idiotypic specificity of the antigen receptor on B cells and T cells is very similar (*Ramseier* and *Lindenmann,* 1969, 1972, 1977; *Binz* and *Wigzell,* 1975, 1977; *Eichmann* and *Rajewski,* 1975; *Rajewsky* and *Eichmann,* 1977). If so, then it seems unlikely that the T and B cell receptors differ vastly. This concept fits best with the idea that the antigen receptor on T cells is probably a separate entity, distinct from the self-interaction mechanism (*Doherty* et al., 1976b; *Janeway* et al., 1976; *Katz,* 1977).

Most of the following experiments address the problem of the antigenic entity recognized by T cells; i.e., whether evidence for neoantigens such as a complex of self and X, modified self, or modified X can be demonstrated.

5. Minor Alloantigens as T Cell Targets. Cytotoxic T cells that are specific for minor alloantigens (*Bevan,* 1975a, b, 1976, 1977a; *Gordon* et al., 1975, 1976; *Simpson* and *Gordon,* 1977) are restricted by *H-2K* or *D.* Although the idea that a complex of self plus minor alloantigen constitutes the single neoantigen is favored by some investigators, it is more readily envisaged that the two cell-surface molecules of self and alloantigen constitute distinct antigenic entities.

6. Relative Susceptibility of Infected Targets to Alloreactive T Cells. Virus-infected cells or cells expressing minor alloantigens seem to be as susceptible to lysis by alloreactive T cells as "normal" target cells (*Koszinowski* and *Ertl,* 1975a, b; *Gardner* et al., 1975; *Zinkernagel* and *Doherty,* 1977a, b) Possibly, cell surface structures that are targets for alloreactive killing (at least some) remain unmodified. If so, either virus infection or minor alloantigen fail to "modify" all self-markers, or the target for alloreactivity is different from the relevant self-marker. However, most evidence suggests that the target for alloreactivity and for self-recognition are identical or very similar and closely linked. Any quantitative effects that suggest modification of self-markers by virus infection can be explained by the shutdown of host-cell protein synthesis after virus infection (*Hecht* and *Summers,* 1972; *Koszinowski* and *Ertl,* 1975a, b). Since we know neither the quantitative antigenic requirements for cytotoxic T cells to act nor the relative frequency of antigens on the target cells, these data cannot argue strongly for or against any of the models.

7. Anti H-2 Antisera Block Antigen Specific Interactions. Rosenthal and *Shevach* were the first to use anti-H-2 antiserum to block antigen-specific T cell proliferation of guinea pigs (*Rosenthal* and *Shevach*, 1973; *Shevach* and *Rosenthal*, 1973). Subsequently, the same sort of blocking was shown for virus-specific or hapten-specific cytotoxic T cell interactions (*Germain* et al., 1975; *Schrader* et al., 1975). These results were complemented only incompletely by the demonstration that killing of the appropriate target cells was sometimes blocked by antiviral or antihapten antibodies (*Gardner* et al., 1974b; *Koszinowski* and *Ertl,* 1975a, b; *Burakoff* et al., 1976a). This pattern may be interpreted to indicate that *H-2* structures form part of the antigenic determinant to which killer cells respond, but simple steric hindrance obviously can not be excluded as the cause of such blocking. This is true even in a linked dual recognition model since the two postulated antigenic entities probably could not lie very far apart. Serologic evidence for a neoantigen in the virus or minor alloantigen systems is lacking. Therefore, positive proof of a serologically definable neoantigen that can block cytotoxicity would certainly support a single neoantigen hypothesis.

8. Evidence for Complex Antigen Formation. Schrader and *Edelman* have attempted to show that virally induced cell-surface antigens co-cap with *H-2K* and *D* determinants on the cell surface (*Schrader* et al., 1975). This result has been interpreted to support the concept that viral antigen and *K* or *D* form a complex. The analysis is however subject to the complication that many of the alloantisera are contaminated with anti-C-type viral antibodies. Although the authors made great efforts to prove that this was not the case with their reagents, a degree of uncertainty remains (*Henning* et al., 1976). The study of *Formann* et al. (1977) is probably the best evidence that, at least in the TNP model, a neoantigen is the crucial determinant for cytotoxic T cells. More recently, *Geib* et al. (1977) demonstrated that the H-Y-alloantigen and *K* and *D* determinants did not co-cap which would be compatible with the idea that most of the self-markers and H-Ys are not complexed. This is certainly no proof for dual recognition, since one can argue that only a few complexed self-H-Y are sufficient to be detectable by T cells even though not apparent in an insensitive test such as co-capping. There is a strong precedence that T killer cells need only few target antigens, since even after complete capping of *H-2* antigens with anti-*H-2* sera, sufficient *H-2* determinant remained for alloreactive killer cells to be active (*Stulting* et al., 1976).

More recently, several groups (*Schrader* and *Edelman,* 1977, *Koszinowski* et al., 1977b) have shown that (noninfectious) viral antigens (Sendai-virus) can, when adsorbed to target cells, render them susceptible to T cell-mediated lysis. This interesting observation demonstrates that the relevant antigen can come from outside the cell and has not necessarily to come from within. Furthermore, in this experimental situation molecular modification of *K* or *D* products by virally induced processes or, alternatively, host-specific modification of viral antigen within the host cell apparently does not occur. However, since Sendai has a great potential for fusion, one cannot fully exclude the possibility that such processes occur during the fusion process on the cell membrane. Nevertheless, these experiments and similar results obtained with vesicular stomatitis virus (*Zinkernagel*, unpublished) and vaccinia virus (*S. Hapel*, personal communication) suggest that viral antigen may be recognized directly as such on infected cells. Whether this actually is interpreted to favor the complex antigen idea implying a simple T cell receptor or the concept of two distinct antigenic entities (self plus X) that are recognized by two observed receptors, seems arbitrary. *Bubbers* and *Lilly* (1977) demonstrated that Friend leukemia virus grown in *H-2^b* cells incorporated serologically detectable *D^b* specificities but not *K^b*. This correlates selectively with the fact that virus-specific killer T cells are generated against the *D^b*-end preferentially. This might suggest that *K* and *D* associate with viral antigens, but does not directly answer the question of the nature of the antigenic moiety seen by T cells.

9. Derepression — Repression of H-2 Antigens. The appearance of neoantigens after virus infection of target cells have been described repeatedly (*Boyse* and *Old*, 1969; *Invernizzi* and *Parmiani*, 1975; *Garrido* et al., 1976). Although most examples were found with tumor-associated viruses, one report (*Garrido* and *Festenstein*, 1976) showed that unexpected alloantigens can be detected by serological means after infection with vaccinia virus. These phenomena were interpreted to suggest that *H-2* antigen expression is not allelic but rather a function of regulator genes controlling expression of multigenically coded H-2 antigens (*Bodmer*, 1973). If viruses can deregulate expression of H-2 antigens, such derepressed specificities may in fact constitute the target antigens. Attempts to find such repression-derepression of alloantigens or self-markers by infecting cells with several viruses have failed when, instead of serologic methods, susceptibility to alloreactive T cells was used as an assay method (*Zinkernagel* et al., 1977b). This would suggest that repression-derepression of *H-2* genes could not readily explain the phenomena of *H-2*-restricted virus-specific T cell cytotoxicity.

10. Cross reactivity of Syngeneic Antigen-Specific Cytotoxic T Cells with Unrelated Antigens or with Alloantigen. The unique neoantigen idea is best applicable to alloantigens. If antialloantigen activity is mediated by a single recognition structure the chance that syngeneic cytotoxicity against antigen X is also specific for an alloantigen B might be considerable. This is particularly to be expected if the T cell receptor is somehow restricted in favor of allorecognition. In the TNP model such cross reactivity has been found extensively; this point has been discussed in an earlier section (*Lemonnier* et al., 1977; *Shearer* et al., 1976). More recently however, *Bevan* has shown that one can select cytotoxic T cells directed against minor alloantigens to find a subpopulation that kills not only H-2-compatible targets expressing this minor alloantigen but also kills one particular foreign H-2 type not expressing the specific, minor alloantigens (*Bevan*, 1977b). This finding is different from the TNP findings, in that the frequency of alloreactive cross reaction is much smaller. Although both findings can be used to support the neoantigen idea (*Matzinger* and *Bevan*, 1977; *Bevan*, 1977b), one can also explain the latter results with a dual recognition model. If allorecognition differs from antigenrecognition for reasons discussed in other sections in that, on target cells the excitable cell-differentiation trigger is hit directly by the recognition arm that is used for antigen X recognition in syngeneic sensitizations, then a certain frequency of alloreactive cells is predicted. The fact that TNP-sensitization creates a large activity against alloantigens, where as cytotoxic T cells generated against minor alloantigens lyse only few allogeneic targets is compatible with the view that the rough chemical modification creates many different modifications of K and D. Similarly, there are so many minor alloantigens that the chance of coming within a binding affinity range against foreign alloantigens that is still low, but high enough to trigger K or D products of cells for efficient lysis to occur, is a real possibility.

11. The TNP Model. The results of the TNP model where T cells are sensitized *in vitro* against TNP coupled to syngeneic target cells indicate that these cytotoxic T cells react to a neoantigen (*Shearer,* 1974; *Shearer* et al., 1976; *Burakoff* et al., 1976a, b; *Lemonnier* et al., 1977). The reason for this contention is as follows: First, TNP-modified target cells are lysed only when conditions are used in which TNP is demonstrably coupled to the target cells' *H-2K* or *D* products (*Forman* et al., 1977a). Second, killer T cells that are activated against syngeneic TNP-modified cells are cross reactive against unmodified, i.e., normal, allogeneic target cells (*Lemonnier* et al., 1977; *Shearer* et al., 1976). This relatively high degree of cross reactivity is unique to the models using chemically coupled antigens since at least in the virus models no cross reactivity between $H\text{-}2^b$ or $H\text{-}2^d$-restricted virus-specific T cells has been discovered that would exceed a level of 3% and in most experiments is at or below the 1% level (*Doherty* and *Zinkernagel,* 1976; *Zinkernagel* et al., 1977b). Third, *Sprent, Wilson* and coworkers (*Wilson* et al., 1977) have shown that lymphocytes that have been made unresponsive against an alloantigen by *in vitro* filtering of parental A lymphocytes through irradiated $A \times B$ F_1 recipients with collection of filtered T cells from the thoracic duct 3–18 h after transfer, could not react against B but could be sensitized against TNP-modified B in mixed lymphocyte cultures. More recently, *Forman* et al. (1977b) showed that mice $H\text{-}2^a$ that were rendered tolerant by injection of $(H\text{-}2^a \times H\text{-}2^b)F_1$ lymphoid cells at birth could generate cytotoxic T cells against both TNP-$H\text{-}2^a$ as well as TNP-$H\text{-}2^b$ targets when stimulated appropriately *in vitro*. Both results are not compatible with a dual recognition model and support for TNP the single neoantigen recognition model. The data from *Thomas* and *Shevach* (1977) agree but the preliminary data from *Schmitt-Verhulst* and *Shearer* do not (personal communication). Using both BUDR plus light to eliminate alloreactive lymphocytes, the latter investigators failed to sensitize lymphocytes specifically against TNP-modified targets possessing the tolerated H-2 haplotype. This result is supported by experiments with chimeras similar to those mentioned in section 3 (*Zinkernagel,* et al., 1978c). Lymphocytes from irradiation chimeras of the type $(H\text{-}2^a \times H\text{-}2^b)$ $F_1 \rightarrow (H\text{-}2^a \times H\text{-}2^c)F_1$ were stimulated with TNP-modified spleen cells from $(H\text{-}2^b \times H\text{-}2^c)F_1$ and then tested for cytotoxic activity; only TNP $H\text{-}2^c$ targets could be lysed, but not TNP-$H\text{-}2^b$. Fourth, if TNP is coupled to target-cell surfaces by means of a spacer instead of directly, killer cells that had been sensitized against TNP coupled directly to syngeneic stimulator cells do not lyse TNP-spacer-modified targets. This result is compatible with the presence of a neoantigen, but could also indicate that the two linked T cell-recognition arms can span only very small distances (*Rehn* et al., 1976).

These data are in support of the view that TNP-self behaves at least in part like an alloantigen and that the rules of alloreactivity may apply directly to this model.

IV. *H-2* Antigens as Receptors for Differentiation Signals

A. Introduction

Why have *H-2* antigens evolved? Many speculations exist (*Thomas*, 1959; *Snell*, 1968; *Burnet*, 1971, 1973; *Jerne*, 1971; *Bodmer*, 1972; *Amos* et al., 1973; *Shreffler* and *David*, 1975; *Klein, 1975; Klein*, 1976). There is some consensus that these cell-surface markers probably evolved before immune systems, e.g., self-recognition has been shown in plants and in coelenteratae (*Hildemann*, 1974; *Theodor*, 1970; *Burnet*, 1971). In fact, some of the striking characteristics of major transplantation antigens are already found in these systems. For example, except for sexual propagation, the contact of two foreign members of colonial tunicates or coelenterates results in necrosis of interacting cells. As in higher vertebrates, related foreigners are recognized much more readily than vastly distinct foreigners. Also as best exemplified in the fecundation process in plants, such self-recognition phenomena may initiate or block differentiation processes.

From this point of view, the *H-2*-restriction of T cell activities resembles these very ancient recognition processes. Particularly if the dual recognition model is applicable, the postulated self-recognition event may well be a very primitive recognition and/or cell-interaction mechanism very much like that found in plants and colony-forming marine animals (*Zinkernagel* et al., 1977c; *Katz*, 1977). As mentioned earlier, it is unclear, whether self-recognition in a dual recognition model operates via a complementing interaction or a like-like self interaction. If the first alternative is true one could postulate that immune recognition derives directly from this ancient mechanism of self-recognition. If the second alternative applies, immune recognition may have evolved quite independently. The postulates that self-recognition moieties may be coded within *H-2* or outside of *H-2* and that the variable regions of immuneglobulins are not linked to *H-2* leave all possibilities open as yet.

B. Allorecognition

Several hypotheses dealing with self-recognition, allorecognition and immune recognition of any antigenic determinants have been formulated (e.g., *Jerne,*

1971; *Doherty* et al., 1976b; *Janeway* et al., 1976; *Zinkernagel* and *Doherty,* 1977b). The main question is whether there is a genetic functional restriction and thus interdependence (and of what degree) between the H-antigens and the *V*-gene region products of the germ line or whether these two gene systems and their products are completely independent.

The argument in favor of some sort of relationship between the two gene systems stems from the puzzling finding of the high frequency of lymphocytes that can "react" immunologically with a species' major transplantation antigenic system (*Simonson,* 1962, 1974; *Jerne,* 1971; *Elkins,* 1971; *Lafferty* and *Cunningham,* 1975; *Bodmer,* 1972). This phenomenon is still unexplained but remains a constant challenge. A simplistic view, which tries to avoid genetic linkage of the *H* and the *V*-gene systems, is that this apparent preoccupation of *V*-gene products with H-antigens is purely "phenotypic", i.e., due to the unique properties of the H-2 antigens recognized but not to some genetic linkage. The unique property of these cell-surface markers that distinguishes them from most other antigens is that *H*-markers are receptors for signals that regulate differentiation and generation of immune effector cells. The greater efficiency with which these processes are triggered directly in the essentially abnormal allogeneic confrontations as compared to the antigen-induced "syngeneic" reaction may thus be explained by the fact that these receptors and the recognized antigen are *identical.* Because of this "abnormalcy", allogeneic stimulator cells provide an immunological trigger to the responder cells (*Lafferty* and *Cunningham,* 1975).

Furthermore, the chance that syngeneic associative recognition can trigger antigen-specific differentiation is much smaller, because of the low frequency of the antigen-specific receptor. Actually, the same small number of receptors may exist for alloantigen if they were conventional as an antigens; however, the conditions for triggering alloreactive cells, as compared with conventional antigens, may be different because of the stimulatory capacity of the allogeneic target cells. Also, at the effector level of a cytotoxic T cell directed against alloantigens, a binding activity of one hundredth of that needed for conventional antigens may be sufficient to lyse the target because the immunologic receptor hits the weak spot on the target cell directly. A very striking example of such abnormal "induction" of immunologic responses is the allogeneic effect. Here, B cells that are confronted with antigen are triggered to differentiate to plaque-forming cells not by syngeneic helper T cells but by alloreactive T cells that probably can trigger the very same differentiation signal-receptors on B cells in an "abnormal" fashion (*Katz,* 1972). In brief, we reason that the very involvement of *H-2* antigens in syngeneic reactions is the reason for the tremendous triggering potential of alloantigens on lymphoid cells because they not only present antigens but also simultaneously stimulate the responding lymphocytes in an abnormal way via the *H-2*-coded receptors for immunologic differentiation; this applies to the induction as well as to the effector level. Therefore, the unusually great number of alloreactive lymphocytes can be explained by the unique stimulatory capacity as well as unique nature of the target antigen and there is no absolute need to postulate a linkage between the respective sets of *V* and *H* genes.

C. Evidence for the Function of *H-2* Coded Antigen as Receptors of Differentiation Signals

The *H-2* gene complex is located on the 17th chromosome. This chromosome also contains other important gene complexes that are thought to be involved in differentiation processes also. For example the *T/t* locus codes for gene products that regulate differentiation processes very early during embryogenesis as summarized excellently by *Hammerberg* and *Klein* (1975) *Klein* (1975), *Artzt* and *Bennet* (1975) and *Jacob* (1977). Another locus (*Tla*) codes for the TL (thymus leukemia) antigens first discovered by *Boyse* and *Old* in 1963 (*Boyse* and *Old*, 1969). The TL alloantigen system consists of a group of antigens that can be detected on normal thymocytes at various stages of differentiation or on leukemia cells. Apparently, both gene complexes code for structural as well as regulatory genes, as is true for the *H-2* gene complex. As mentioned earlier, it is unclear as yet what their exact mutual relationship is. Furthermore *H-2* is linked to both the *T/t* locus and the *Tla* locus. A strong linkage desequilibrium exists between *T/t* and *H-2*, and the expression of heir respective gene products is mutually dependent during differentiation. Similarly, the expressions of TL antigen and of some *H-2* products are linked. An additional relationship exists between *H-2* and TL antigens in that their respective expressions may be modified by infection with tumor-associated viruses. *Boyse* and *Old* suggested that C-type RNA leukemia viral genes themselves may be integrated at the *Tla* locus and the expression of viral antigen may be coregulated with differentiation processes of thymocytes (*Boyse* and *Old,* 1969). Similarly, several investigators (*Invernizzi* and *Parmiania,* 1975; *Garrido* and *Festenstein,* 1976, *Garrido* et al., 1976) found that in certain cell lines infected with tumor-associated viruses, the phenotypic expression of serologically detectable *H-2* antigens changed; in cells infected with acute viruses (*Zinkernagel* and *Klein,* 1977b), such changes could not be detected by alloreactive cytotoxic T cells. To what extent parts of these genetic loci and complexes including *H-2* can be regarded as integrated viral information or the respective gene products as "footprints" of viruses is unclear at the moment; however, the hypothesis of viral integration is attractive and may explain at least some aspects of the functional immunologic association between acute viruses and some products of *H-2* (*Huebner* and *Todaro,* 1969; *Todaro,* 1975; *Del Villano* et al., 1975; *Zinkernagel,* 1977). Such a hypothesis could also explain why tumor-associated viruses, but not acute viruses, may be able to disturb such well-established symbioses and escape immune surveillance (this term is used in a much broader sense as proposed by *Thomas* and *Burnett* to include all immunoreactivity to self-plus-foreign not only to tumor antigens), because they successfully use and exploit self-recognition markers for this purpose.

1. K and D as Receptors for Lytic Signals. T cells that are specific for the serologically defined products of *H-2K* or *H-2D*, be it in allogeneic confrontation of the transplantation reaction or in syngeneic ractions against viruses, destroy the target cell. Cell lysis seems to be caused by dysregulation of the osmolar balance between extracellular and intracellular compartments. Whether this occurs by shut off of one of the ion pumps to establish ion gradients or by actual overshooting activity of such pumps is unclear. One may suggest that K and D products are linked to regulators of the osmolar balance or are actively involved themselves. How these various regulators, ion-pumps and K or D structure, are actually linked is unclear; also, whether there exists a system such as the actin-myosin network that interconnects similar structures and coordinates processes is pure speculation (*Singer,* 1977).

From a teleologic point of view, target cell destruction as a means of protecting the host from potential harm may be functional in the following examples: Elimination of cells that have escaped normal controls over the rate of multiplication and anatomic restriction ("tumor cells"); here the cell itself is the target of destruction. Alternatively, a cell may be the target of destruction if it harbors parasites that hide inside the cell wall. However, this latter mechanism seems to exist only to eliminate intracellular parasites, which themselves destroy cells when growing in them. Acute viruses seem to be an excellent example of such parasites, since many of them shut down host protein synthesis and thus kill the host cell functionally. However, during infection, these viruses induce cell-surface changes very rapidly and the host cell may thus be susceptible to immuno-logic attack (*Ada* et al., 1976; *Koszinowski* et al., 1977b). In fact, virus-specific cytolytic T cells have been shown to lyse target cells well before assembly of viral progeny, at least in the case of vaccinia virus. Once viral progeny do assemble, host-cell lysis no longer affects the newly formed viruses, and the infected host can overcome viral invasion by cytotoxic T cells only after infection of surrounding cells by these progeny. However, noncytolytic viruses (such as lymphocytic choriomeningitis virus — LCMV) that do not halt host-cell functions, no longer release viral progeny after host cells die. Thus, viruses can be efficiently eliminated by target-cell lysis if it occurs during the eclipse phase of the virus infection, i.e., after penetration and uncoating and before reassembly of infectious progeny. It has to be kept in mind that many other antiviral immune mechanisms exist, such as neutralizing antibodies, immune interferon, etc. A second fact favors the idea that K and D products are receptors of T cell-mediated cell destruction. This fact is the ubiquitous distribution of K and D structures. So far, K and D cell-surface markers have been demonstrated on all nucleated cells, although the quantities vary (*Klein,* 1975). For example, neurons possess about a 1000 times less H-$2K$ or D structures than lymphocytes. Since, viruses can actively infect phagocytic as well as nonpha-gocytic cells, the potential of different viruses to infect particular kinds of cells would require that the immune system use an ubiquitous self-marker if cell destruction is one of its means of eliminating virus. Thus K and D structures appear to fit both these requirements: receptors for lytic signals and ubiquitous distribution.

Why have viruses not escaped immune surveillance? Obviously viruses can adapt many times more quickly than their host, because of the vast difference in generation time. One explanation may be that virus life cycles and host defense mechanisms have established an overall optimal steady state condition. If cytopathic viruses lysed too many cells too quickly, hosts would die before they could efficiently propagate the virus. Similarly, too potent an immune response would result in virus elimination without propagation. The actual kinetics of both virus spread in a host and throughout a population on one hand and of the immune response on the other hand is a compromise for mutual survival.

During the last few years, several groups of investigators have shown that under certain experimental conditions that include *in vivo* priming and *in vitro* secondary restimulation, it is possible to sensitize cytotoxic T cells that are specific for *H-2* region determinants (*Wagner* et al., 1975; *Nabholz* et al., 1975). More recently it was found that *I* region determinant-directed killing was not *K* or *D* restricted (*Klein* et al., 1977; *Billings* et al., 1977). These results may indicate that the functional roles of *K* and *D* versus *I*-coded structures may be less absolute than thought so far. However, it is interesting that virus specific cytotoxic T cell activity that was generated *in vivo* is, as far as can be detected in ^{51}Cr-release assays or in adoptive transfer assays, exclusively directed against *K* and *D* (*Blanden* et al., 1975a; *Shearer* et al., 1976; *Blanden* et al., 1975b; *Kees* and *Blanden*, 1976; *Doherty* et al., 1976a; *Zinkernagel* and *Welsh*, 1976). Thus, the efficiency with which *K, D*-specific cytotoxic syngenic T cells are generated *in vivo* as compared to *I*-directed killer T cells is of the order of at least 100 fold better; consequently the biologic role of *I*-specific cytotoxicity is unclear as yet.

2. H-2I Coded Structures as Receptors for Nonlytic Differentiation Signals. Except for the examples mentioned, in general, *H-2I*-specific T cell functions seem to be nonlytic. Thus, T cells involved in T-B cell cooperation (*Kindred* and *Shreffler*, 1972; *Katz* et al., 1973a, b; *Katz* and *Benacerraf*, 1975; *Katz*, 1977), in mediation of delayed-type hypersensitivity (DTH) (*Miller* et al., 1975; *Miller* et al., 1976), in proliferative responses (*Rosenthal* and *Shevach*, 1973; *Paul* et al., 1977), in the production of lymphokines such as migration inhibition factor (MIF), or in protection against intracellular bacteria such as *Listeria mono-cytogenes* (*Zinkernagel* et al., 1977c) all produce the differentiation of cell-specific special functions that lead to "neutralization" (i.e., destruction, and/or biologic inactivation) of foreign, potentially harmful antigens. Such neutralization may occur via several mechanisms, e.g., by antibodies alone or combined with complement or by release of chemotactic substances that attract phagocytic cells, which when activated digest engulfed antigens more efficiently. The mechanism of cell-mediated protection against intracellular bacteria such as *L. monocytogenes* may be an excellent representative for antigens that are handled via *I*-coded self-markers. *Mackaness* and his school have elaborated the immune mechanisms that lead to the protection against intracellular bacteria (*Mackaness*, 1962, 1964, 1969; *Blanden* et al., 1969). T cells are specifically triggered to release lymphokines that activate macrophages to nonspecifically increased bactericidal capacity (*Lane* and *Unanue*, 1972; *Blanden* and *Langman*, 1972; *North*, 1973).

It is obvious that lysis of the phagocytic cells would not result in inactivation or destruction of these intracellular parasites; similarly toxins that have become cell associated are not neutralized by the cell's destruction. In neither case do bacteria nor toxins lose their ability to replicate or act as toxins when entering cells. Furthermore, these bacterial antigens are, distinct from viruses (or tumor antigens) not "active" and associate predominantly, if not exclusively, with phagocytic cells to become immunogenic for T cells.

Therefore, again from a teleologic point of view, the self-markers that are involved in handling these "inactive" or inert antigens immunologically, should fulfill two conditions; one, they should be expressed selectively on cells that phagocytize these antigens or are otherwise involved in disposing of them immunologically (macrophages, B cells, T cells, etc.), and two, they should represent receptors for signals that trigger differentiation of cell-specific functions whereby inert antigens are effectively neutralized. On macrophages such a differentiative step could lead to the activation of enzyme systems that improve their phagocytic and digestive capacity; on B cells the signal may initiate immunoglobulin production and/or the switch from IgM to IgG production; on T cells it may induce proliferation, etc.

Thus, I-coded structures may be partially cell-specific, cell-surface structures that function as receptors for final differentiation signals.

V. Concluding Remarks

All T cell-mediated functions that have been tested so far in mice obey the rule of H-2 restriction. This restriction fundamentally distinguishes the recognition mechanism of T cells from that of B cells or of the B cell product, antibodies. The H-2 restriction apparently generalizes the rule of associative recognition that was first established for T helper cells in the hapten-carrier system, i.e., self-determinants fulfill a carrier task. The fact that B cells do not recognize self-plus-X, but respond to X exclusively — obviously an easily excitable system — explains why deviations arise so often. The encounter of autoantibodies of various kinds in normal but also in many pathologic situations is in great contrast to the fact that no T cell activities have been detected that are directed against normal cell-surface antigens. In general, all so called cell-mediated autoimmune phenomena are most likely T cell activities against self-plus-X, X probably often being viral antigens; some examples are autoaggressive hepatitis in humans or the T cell-mediated immunopathology caused by lymphocytic choriomeningitis virus (Reviewed in *Cole* and *Nathanson*, 1975). Thus, associative selfplux-X recognition is a most efficient self-control device for cell-mediated immunologic lymphocyte interactions. Such an extremely restricting mechanism is much more efficiently regulated via two linked receptors than via a single T cell receptor because single receptors are subject only to clonal selection but are not also regulated by genic and/or allelic exclusion mechanism, as dual receptors would be.

Such complex regulation allows self-recognition to be expressed during differentiation. As has been proposed, when these self-recognizers are no longer expressed in the anti-X repetoire by allelic and/or genic exclusion then autoaggression in the strict sense cannot arise (*Langman*, 1977).

However, since we do not know whether self-recognition and recognition of X involve the same basic receptors, this proposal is only a speculation.

For example, other hypotheses evoke different mechanisms for self-recognition (namely glycosyltransferase-type interactions (*Rothenberg, 1976; Blanden* et al., 1976a) than for recognition of X. In this case, regulation of autoimmune receptors arising against self in the anti-X compartment would be as likely for T cells as for B cells.

From this point of view, the model advocating dual receptors, which may be linked and are subject to allelic and/or genic exclusion, would be the more appealing characterization of T cell recognition. Unfortunately, positive proof of either dual receptors or of a single receptor for a complex antigen for altered self or for self-plus-antigen X is still lacking, as becomes obvious from the experiments that have been reviewed here.

Irrespective of whether single or dual recognition is the more probable mechanism of T cell interaction, the experimental results discussed here suggest that histocompatibility and immunity are interrelated and that major histocompatibility antigens function as receptors for partially cell-specific immunologic differentiation signals.

Acknowledgements. I thank Dr. F.J. Dixon for support, Dr.D.H. Katz for helpful comments and Ms. Beth Sinclair, Judy Henneke and Phyllis Minick for their excellent assistance in preparing the manuscript. Part of the experimental work was supported by USPHSG A1 07007 and A1 13779. This is publication no. 41 of the Department of Cellular and Developmental Immunology and publication no. 1280 of the Departments of Immunology, and was completed in July 1977.

References

Ada, G.L., Jackson, D.C., Blanden, R.V., Tha Hla, R., Bowern, N.A.: Changes in the surface of virus-infected cells recognized by cytotoxic T cells. I. Minimal requirements for lysis of ectromelia-infected P-815 cells. Scand. J. Immunol. *5*, 23–30 (1976)

Amos, D.B., Inou, T, Rowland, D.T.: Human histocompatibility and susceptibility to disease. Science *182*, 183 (1973)

Artzt, K., Bennett, D.: Analogies between embryonic (T/t) antigens and adult major histocompatibility (H-2) antigens. Nature (London) *256*, 545–547 (1975)

Bechtol, K.B., Freed, J.H., Herzenberg, L.A., McDevitt, H.O.: Genetic control of the antibody response to POLY-L(TYR, GLU)-POLY-D, L-ALA-POLY-L-LYS in C3H↔CWB tetraparental mice. J. Exp. Med. *140*, 1660 (1974)

Benacerraf, B., McDevitt, H.O.: Histocompatibility-linked immune response genes. Science *175*, 273 (1972)

Bevan, M.J.: Interaction antigens detected by cytotoxic T cells with the major histocompatibility complex as modifier. Nature (London) *256*, 419 (1975a)

Bevan, M.J.: The major histocompatibility complex determines susceptibility to cytotoxic T cells directed against minor histocompatibility antigens. J. Exp. Med. *142*, 1349 (1975b)

Bevan, M.J.: Cross-priming for a secondary cytotoxic response to minor H antigens with *H-2* congenic cells which do not cross-react in the cytotoxic assay. J. Exp. Med. *143*, 1283 (1976)

Bevan, M.J.: Cytotoxic T-cell response to histocompatibility antigens: the role of H-2. In: Origins of Lymphocyte Diversity, Volume XLI, Cold Spring Harbor Symp. Quant. Biol. (1977a), p. 519

Bevan, M.: Killer cell reactive to altered-self antigens can also be alloreactive. Proc. Natl. Acad. Sci. USA *74*, 2094–2098 (1977b)

Bevan, M.J.: In radiation chimeras host H-2 antigen determine the immune responsiveness of donor cytotoxic cells. Nature (London) *269*, 417 (1977)

Billings, P., Burakoff, S., Dorf, M.E., Benacerraf, B.: Cytotoxic T lymphocytes specific for *I* region determinants do not require interactions with *H-2K* or *D* gene products. J. Exp. Med. *145*, 1387–1392 (1977)

Binz, H., Wigzell, H.: Shared idiotypic determinants on B and T lymphocytes reactive against the same antigenic determinants. I. Demonstration of similar or identical idiotypes on IgG molecules and T-cell receptors with specificity for the same alloantigens. J. Exp. Med. *142*, 197 (1975)

Binz, H., Wigzell, H.: Antigen-binding, idiotypic receptors from T lymphocytes: an analysis of their biochemistry, genetics, and use as immunogens to produce specific immune tolerance. In: Origins of Lymphocyte Diversity, Volume XLI, Cold Spring Harbor Symp. Quant. Biol. (1977) p. 275–284

Blanden, R.V.: T cell response to viral and bacterial infection. Transplant. Rev. *19*, 56 (1974)

Blanden, R.V., Langman, R.E.: Cell-mediated immunity to bacterial infection in the mouse. Thymus-derived cells as effectors of acquired resistance to *Listeria monocytogenes.* Scand. J. Immunol. *1*, 379–391 (1972)

Blanden, R.V., Doherty, P.C., Dunlop, M.B.C., Gardner, I.D., Zinkernagel, R.M., David, C.S.: Genes required for T cell mediated cytotoxicity against virus infected target cells are in the K or D regions of the H-2 gene complex. Nature (London) *254*, 269–270 (1975a)

Blanden, R.V., Bowern, N.A., Pang, T.E., Gardner, I. and *Parish, C.R.:* Effects of thymus-independent (B) cells and the *H-2* gene complex on antiviral function of immune thymus-derived (T) cells. Aust. J. Exp. Biol. Med. Sci. *53*, 187 (1975b)

Blanden, R.V., Dunlop, M.B.C., Doherty, P.C., Kohn, H.I., McKenzie, I.F.C.: Effects of four *H-2K* mutations on virus-induced antigens recognised by cytotoxic T cells. Immunogenetics *3*, 541–548 (1976)

Blanden, R.V., Hapel, A.J., Jackson, D.: Mode of action of Ir genes and the nature of T cell receptors for antigen. Immunochemistry *13*, 179 (1976)

Blanden, R.V., Lefford, M.J., Mackaness, G.B.: The host response to calmetteguerin bacillus infection in mice. J. Exp. Med. *129*, 1079–1107 (1969)

Bodmer, W.F.: Evolutionary significance of the HL-A system. Nature *237*, 139–145 (1972)

Bodmer, W.F.: A new genetic model for allelism at histocompatibility and other complex loci: polymorphism for control of gene expression. Transplant. Proc. *5*, 1471–1475 (1973)

Boehmer, H. von: The cytotoxic immune response against male cells: control by two genes in the murine major histocompatibility complex. Basel Institute of Immunology 1977

Boehmer, H. von, Hass, W.: Cytotoxic T lymphocytes recognize allogeneic tolerated TNP-conjugated cells. Nature (London) *261*, 139–140 (1976)

Boehmer, H. von, Hudson, L., Sprent, J.: Collaboration of histoincompatible T and B lymphocytes using cells from tetraparental bone marrow chimeras. J. Exp. Med. *142*, 989 (1975)

Boyse, E.A., Old, L.J.: Some aspects of normal and abnormal cell surface genetics. Am. Rev. Genetics *3*, 269–290 (1969)

Bryere, E.J., Williams, L.B.: Antigens associated with a tumor virus: rejection of isogenic skin grafts from leukemic mice. Science *146*, 1055 (1964)

Bubbers, J.E., Lilly, F.: Selective incorporation of *H-2* antigenic determinants into Friend virus particles. Nature (London) *266*, 458 (1977)

Burakoff, S.J., Germain, R.N., Benacerraf, B.: Cross-reactive lysis of trinitrophenyl (TNP)-derivatized *H-2* incompatible target cells by cytolytic T lymphocytes generated against syngeneic TNP spleen cells. J. Exp. Med. *144*, 1609 (1976b)

Burakoff, S.J., Germain, R.N., Dorf, M.E., Benacerraf, B.: Inhibition of cellmediated cytolysis of trinitrophenyl-derivatized target cells by alloantisera directed to the products of the K and D loci of the *H-2* complex. Proc. Natl. Acad. Sci. USA *73*, 625 (1976a)

Burnet, F.M.: "Self-recognition" in colonial marine forms and flowering plants in relation to the evolution of immunity. Nature (London) *232*, 230 235 (1971)

Burnet, F.M.: Multiple polymorphism in relation to histocompatibility antigens. Nature (London) *245,* 359–361 (1973)

Cerottini, J.C., Brunner, K.T.: Cell-mediated cytotoxicity, allograft rejection and tumor immunity. Adv. Immunol. *19,* 67 (1974)

Claman, H.N., Chaperon, E.A.: Immunologic complementation between thymus and marrow cells—a model for the two-cell theory of immunocompetence. Transplant. Rev. *1,* 92 (1969)

Claman, H.N., Chaperon, E.A., Triplett, R.F.: Thymus-marrow cell combinations. Synergism in antibody production. Proc. Soc. Exp. Biol. Med. *122,* 1167 (1966)

Cole, G.A. and *Nathanson, N.:* Lymphocytic choriomeningitis: pathogenesis. Prog. Med. Virol. *18,* 94 (1975)

Dausset, J.: Correlation between histocompatibility antigens and susceptibility to illness. Prog. Clin. Immunol. *1,* 183 (1972)

Debré, P., Waltenbaugh, C., Dorf, M.E., Benacerraf, B.: Genetic control of specific immune suppression. III. Mapping of *H-2* complex complementing genes controlling immune suppression by the random copolymer L-glutamic acid^{50}-L-tyrosine^{50} (GT). J. Exp. Med. *144,* 272 (1976)

Del Villano, B.C., Kennel, J.J., Lerner, R.A.: Biological and structural pleomorphism of the oncornavirus envelope glycoprotein gp70. Contemp. Top. Immunobiol. *6,* 16 (1975)

Doherty, P.C., Zinkernagel, R.M.: T-cell mediated immunopathology in viral infections. Transplant. Rev. *19,* 89–120 (1974)

Doherty, P.C., Zinkernagel, R.M.: H-2 compatibility is required for T cell mediated lysis of target cells infected with lymphocytic choriomeningitis virus. J. Exp. Med. *141,* 502–507 (1975a)

Doherty, P.C., Zinkernagel, R.M.: A biological role for the major histocompatibility antigens. Lancet *1,* 1406–1409 (1975b)

Doherty, P.C., Zinkernagel, R.M.: Specific immune lysis of paramyxovirus-infected cells by H-2 compatible thymus-derived lymphocytes. Immunology *31,* 27–32 (1976)

Doherty, P.C., Blanden, R.V., Zinkernagel, R.M.: Specificity of virus-immune effector T cells for H-2K or H-2D compatible interactions: Implications for H-antigen diversity. Transplant. Rev. *29,* 89–124 (1976a)

Doherty, P.C., Götze, D., Trinchieri, G., Zinkernagel, R.M.: Models for regulation of virally modified cells by immune thymus derived lymphocytes. Immunogenetics *3,* 517–524 (1976b)

Dorf, M.E., Benacerraf, B.: Complementation of *H-2*-linked *Ir* genes in the mouse. Proc. Natl. Acad. Sci. USA *72,* 3671 (1975)

Eichmann, K., Rajewsky, K.: Induction of T and B cell immunity by anti-idiotypic antibody. Eur. J. Immunol. *5,* 661 (1975)

Elkins, W.L.: Cellular immunology and the pathogenesis of graft versus host reactions. Prog. Allergy *15,* 78 (1971)

Erb, P., Feldman, M.: The role of macrophages in the generation of T helper cells. II. The genetic control of the macrophage-T-cell interaction for helper cell induction with soluble antigens. J. Exp. Med. *142,* 460 (1975)

Forman, J.: On the role of the *H-2* histocompatibility complex in determining the specificity of cytotoxic effector cells sensitized against syngeneic trinitrophenyl-modified targets. J. Exp. Med. *142,* 403–418 (1975)

Forman, J.: The specificity of thymus-derived T-cells in cell-mediated cytotoxic reactions. Transplant. Rev. *29,* 146–163 (1976)

Forman, J., Klein, J.: Immunogenetic analysis of H-2 mutations. VI. Cross-reactivity in cell-mediated lympholysis between TNP-modified cells from H-2 mutant strains. Immunogenetics *4,* 183 (1977)

Forman, J., Vitetta, E.S.: Absence of H-2 antigens capable of reacting with cytotoxic T cells on a teratoma line expressing a T/t locus antigen. Proc. Natl. Acad. Sci. USA *72,* 3661 (1975)

Forman, J., Vitetta, E.S., Hart, D.A.: Relationship between trinitrophenyl and H-2 antigens on trinitrophenyl-modified spleen cells. II. Correlation between derivatization of H-2 antigens with trinitrophenyl and the ability of trinitrophenylmodified cells to react functionally in the CML assay. J. Immunol. *118,* 803–808 (1977a)

Forman, J., Klein, J. and *Streilein, J.W.:* Spleen cells from animals neonatally tolerant to H-2K^k antigens recognize trinitrophenyl-modified H-2K^k spleen cells. Immunogenetics *5,* 561–567 (1977b)

Gardner, I., Bowern, N.A., Blanden, R.V.: Cell-mediated cytotoxicity against ectromelia virus-infected target cells. I. Specificity and kinetics. Eur. J. Immunol. *4,* 63 (1974a)

Gardner, I., Bowern, N.A., Blanden, R.V.: Cell-mediated cytotoxicity against ectromelia virus-infected target cells. II. Identification of effector cells and analysis of mechanism. Eur. J. Immunol. *4,* 68 (1974b)

Gardner, I., Bowern, N.A., Blanden, R.V.: Cell-mediated cytotoxicity against ectromelia virus-infected target cells. III. Role of the *H-2* gene complex. Eur. J. Immunol. *5,* 122 (1975)

Garrido, F., Festenstein, H.: Further evidence for derepression of H-2 and Ia-like specificities of foreign haplotypes in mouse tumor cell lines. Nature (London) *261,* 705–707 (1976)

Garrido, F., Schirrmacher, V., Festenstein, H.: H-2-like specificities of foreign haplotypes appearing on a mouse sarcoma after vaccinia virus infection. Nature (London) *259,* 228–230 (1976)

Geib, R., Goldburg, E.H., Klein, J.: Membrane bound H-2 and H-Y antigens move independenty of each other. Nature (London) *270,* 252–254 (1977)

Germain, R.M., Dorf, M.E., Benacerraf, B.: Inhibition of T-lymphocyte mediated tumor-specific lysis by alloantisera directed against the H-2 serological specificities of the tumor. J. Exp. Med. *142,* 1023 (1975)

Gordon, R.D., Simpson, E. and *Samelson, L.E.:* In vitro cell-mediated immune responses to the male specific (H-Y) antigen in mice. J. Exp. med. *142,* 1108–1114 (1975)

Gordon, R.D., Mathieson, B.J., Samelson, L.E., Boyse, E.A., Simpson, E.: The effect of allogeneic presensitization on H-Y graft survival and in vitro cell-mediated responses to H-Y antigen. J. Exp. med. *144,* 810–820 (1976)

Goulmy, E., Termijtelen, A., Bradley, B.A., Van Rood, J.J.: Y-antigen killing by T cells of women is restricted by HLA. Nature (London) 266, 544 (1976)

Hammerberg, C., Klein, J.: Linkage disequilibrium between H-2 and t complexes in chromosome 17 of the mouse. Nature (London) *258,* 296–299 (1975)

Hecht, T.T., Summers, D.F.: Effect of vesicular stomatitis virus infection on the histocompatibility antigen of L cells. J. Virol. *10,* 578–584 (1972)

Henning, R., Schrader, J.W., Edelman, G.M.: Antiviral antibodies inhibit the lysis of tumour cells by anti-H-2 sera. Nature (London) *263,* 689–691 (1976)

Hildemann, W.H.: Some new concepts in immunological phylogeny. Nature (London) *250,* 116 (1974)

Huebner, R.J., Todaro, G.J.: Oncogenes of RNA tumor viruses as determinants of cancer. Proc. Natl. Acad. Sci. USA *64,* 1087–1094 (1969)

Invernizzi, G., Parmiani, G.: Tumor-associated transplantation antigens of chemically induced sarcomata cross reacting with allogeneic histocompatibility antigens. Nature (London) *254,* 713–714 (1975)

Jacob, F.: Mouse teratocarcinoma and embryonic antigens. Transplant. Rev. *33,* 3–31 (1977)

Janeway, C.A., Jr., Wigzell, H., Binz, H.: Two different V_H gene products make up the T-cell receptors. Scand. J. Immunol. *5,* 993 (1976)

Jerne, N.K.: The somatic generation of immune recognition. Eur. J. Immunol. *1,* 1 (1971)

Katz, D.H.: The allogeneic effect on immune responses: model for regulatory influences of T lymphocytes on the immune system. Transplant. Rev. *12,* 141 (1972)

Katz, D.H.: The role of the histocompatibility gene complex in lymphocyte differentiation. In: Origins of Lymphocyte Diversity, Vol. XLI, Cold Spring Harbor Symp. Quant. Biol. (1977)

Katz, D.H., Benacerraf, B.: The function and interrelationship of T cell receptors, Ir genes and other histocompatibility gene products. Transplant. Rev. *22,* 1975 (1975)

Katz, D.H., Benacerraf, B.: Editors, In: The Role of Products of the Histocompatibility Gene Complex in Immune Responses. New York: Academic Press 1976

Katz, D.H., Graves, M., Dorf, M.E., Dimuzio, H., Benacerraf, B.: Cell interactions between histoincompatible T and B lymphocytes. VII. Cooperative responses between lymphocytes are controlled by genes in the *I* region of the *H-2* complex. J. Exp. Med. *141*, 263 (1975)

Katz, D.H., Hamaoka, T., Benacerraf, B.: Cell interactions between histoincompatible T and B lymphocytes. II. Failure of physiologic cooperative interactions between T and B lymphocytes from allogeneic donor strains in humoral response to hapten-protein conjugates. J. Exp. Med. *137*, 1405–1418 (1973a)

Katz, D.H., Hamaoka, T., Dorf, M.E., Benacerraf, B.: Cell interactions between histoincompatible T and B lymphocytes. The *H-2* gene complex determines successful physiologic interactions. Proc. Natl. Acad. Sci. USA *70*, 2624 (1973b)

Katz, D.H., Paul, W.E., Goidl, E.A., Benacerraf, B.: Carrier function in antihapten antibody responses. III. Stimulation of antibody synthesis and facilitation of hapten-specific secondary antibody responses by graft-versus-host reactions. J. Exp. Med. *133*, 169–186 (1971)

Kees, U., Blanden, R.V.: A single genetic element in *H-2K* affects mouse T-cell antiviral function in poxvirus infection. J. Exp. Med. *143*, 450 (1976)

Kindred, B., Shreffler, D.C.: H-2 dependence of cooperation between T and B cells *in vivo*. J. Immunol. *109*, 940 (1972)

Klein, J.: Biology of the Mouse Histocompatibility-2 Complex. Heidelberg-Berlin-New York: Springer, 1975

Klein, J.: An attempt at an interpretation of the mouse *H-2* complex, In: Contemporary Topics in Immunobiology. *Weigle, W.O.* (Ed.), New York: Plenum Press 1976, Vol. 5, p. 297

Klein, J., Chiang, C.L., Hauptfeld, V.: Histocompatibility antigens controlled by the *I* region of the murine *H-2* complex. II. K/D region compatibility is not required for *I*-region cell-mediated lymphocytotoxicity. J. Exp. Med. *145*, 450–454 (1977)

Koszinowski, U., Ertl, H.: Lysis mediated by T cells and restricted by H-2 antigen of target cells infected with vaccinia virus. Nature (London) *255*, 552 (1975a)

Koszinowski, U., Ertl, H.: Altered serological and cellular reactivity to H-2 antigens after target cell infection with vaccinia virus. Nature (London) *257*, 596 (1975b)

Koszinowski, U., Ertl, H., Wekerle, H., Thomssen, R.: Recognition of alterations induced by early vaccinia surface antigens and dependence of virus-specific lysis on H-2 antigen concentration on target cells. In: Origins of Lymphocyte Diversity, Volume XLI, Cold Spring Harbor Symp. Quant. Biol. (1977a), p. 529

Koszinowski, U., Gething, M.J., Waterfield, M.: T-cell cytotoxicity in the absence of viral protein synthesis in target cells. Nature (London) *267*, 160 (1977b)

Lafferty, K.J., Cunningham, A.J.: A new analysis of allogeneic interactions. Aust. J. Exp. Viol. Med. Sci. *53*, 27 (1975)

Lane, F.C., Unanue, E.R.: Requirement of thymus (T) lymphocytes for resistance to listeriosis. J. Exp. Med. *135*, 1104–1112 (1972)

Langman, R.E.: The Role of the Major Histocompatibility Complex in Immunity: A new concept in the functioning of a cell-mediated immune system. Rev. Physiol. Biochem. Pharmacol.

Lawrence, H.S.: Homograft sensitivity. Physiol. Rev. *39*, 811 (1959)

Lawrence, H.S.: Transfer factor in cellular immunity. In: The Harvey Lectures, Series 68 (1974)

Lemonnier, F., Burakoff, S.J., Germain, R.N., Benacerraf, B.: Cytolytic thymus-derived lymphocytes specific for allogeneic stimulator cells crossreact with chemically modified syngeneic cells. Proc. Natl. Acad. Sci. USA *74*, 1229 (1977)

Lilly, F.: The influence of histocompatibility-2 type on response to the Friend leukemia virus in mice. J. Exp. Med. *127*, 665 (1968)

Lilly, F., Pincus, T.: Genetic control of murine viral leukemogenesis. Adv. Cancer Res. *17*, 231–279 (1973)

Lilly, F., Boyse, E.A., Old, L.J.: Genetic basis of susceptibility to viral leukemogenesis. Lancet 2, 1207–1209 (1964)

Lindenmann, J., Klein. P.A.: Immunological aspects of viral oncolysis. In: Recent Results in Cancer Research. Springer Publishing Co. Inc. New York 9, 1–85 (1967)

Lundstedt, C.: Interaction between antigenically different cells. Virus-induced cytotoxicity by immune lymphoid cells in vitro. Acta Pathol. Microbiol. Scand. *75*, 139–152 (1969)

Mackaness, G.B.: Cellular resistance to infection. J. Exp. Med. *116,* 381–406 (1962)

Mackaness, G.B.: The immunological basis of acquired cellular resistance. J. Exp. Med. *120,* 105–120 (1964)

Mackaness, G.B.: The influence of immunologically committed lymphoid cells on macrophage activity in vivo. J. Exp. Med. *129,* 973–992 (1969)

Marker, O., Volker, M.: Studies on cell-mediated immunity to lymphocytic choriomeningitis virus in mice. J. Exp. Med. *137,* 1511 (1973)

Marshak, A., Doherty, P.C., Wilson, D.B.: The control of specificity of cytotoxic lymphocytes by the major histocompatibility complex (Ag-B) in rats and identification of a new alloantigen system showing no Ag-B restriction. J. Exp. Med. *146,* 1773–1790 (1977)

Matzinger, P., Bevan, M.J.: Hypothesis: Why do so many lymphocytes respond to major histocompatibility antigens? Cell Immunol. *29,* 1 (1977)

Miller, J.F.A.P., Mitchell, G.F.: Cell to cell interaction in the immune response. I. Hemolysin-forming cells in neonatally thymectomized mice reconstituted with thymus or thoracic duct lymphocytes. J. Exp. Med. *128,* 801 (1968)

Miller, J.F.A.P., Vadas, M.A., Whitelaw, A., Gamble, J.: *H-2* gene complex restricts transfer of delayed-type hypersensitivity in mice. Proc. Natl. Acad. Sci. USA *72,* 5095 (1975)

Miller, J.F.A.P., Vadas, M.A., Whitelaw, A., Gamble, J.: Role of major histocompatibility complex gene products in delayed-type hypersensitivity. Proc. Natl. Acad. Sci. USA *73,* 2486–2490 (1976)

Mitchell, G.F., Miller, J.F.A.P.: Cell to cell interaction in the immune response. II. The source of hemolysin-forming cells in irradiated mice given bone marrow and thymus or thoracic duct lymphocytes. J. Exp. Med. *128,* 821 (1968)

Mitchison, N.A.: Passive transfer of transplantation immunity. Proc. R. Soc. Lond. Ser. B. *142,* 72–87 (1954)

Mitchison, N.A.: The carrier effect in the secondary response to hapten-protein conjugates. I. Measurement of the effect with transferred cells and objections to the local environment hypothesis. Eur. J. Immunol. *1,* 10 (1971 a)

Mitchison, N.A.: The carrier effect in the secondary response to hapten-protein conjugates. II. Cellular cooperation. Eur. J. Immunol. *1,* 18 (1971 b)

Morris, P.J.: Histocompatibility systems immune response and disease in man. Contemp. Top. Immunobiol. *3,* 141 (1974)

Munro, A.J., Bright, S.: Products of the major histocompatibility complex and their relationship to the immune response. Nature (London) *264,* 145–152 (1976)

Munro, A.J., Taussig, M.J.: Two genes in the major histocompatibility complex control immune response. Nature (London) *256,* 103 (1975)

Nabholz, M., Young, H., Rijnbeek, A., Boccardo, R., David, C.S., Meo, T., Miggiano, V., Shreffler, D.C.: I-region associated determinants: expression on mitogen stimulated lymphocytes and detection by cytotoxic T cells. Eur. J. Immunol. *5,* 594 (1975)

North, R.J.: Importance of thymus-derived lymphocytes in cell-mediated immunity to infection. Cell. Immunol. *7,* 166–176 (1973)

Oldstone, M.B.A., Dixon, F.J.: Tissue injury in lymphocytic choriomeningitis viral infection: virus-induced immunologically specific release of cytotoxic factor from immune lymphoid cells. Virology *112,* 805 (1970)

Paul, W.E., Benacerraf, B.: Functional specificity of thymus-dependent lymphocytes. Science *195,* 1293 (1977)

Paul, W.E., Shevach, E.M., Thomas, D.W., Pickeral, S.F., Rosenthal, A.S.: Genetic restriction in T-lymphocyte activation by antigen-pulsed peritoneal exudate cells. In: Origins of Lymphocyte Diversity, Volume XLI, Cold Spring Harbor Symp. Quant. Biol. (1977a)

Pfizenmaier, K., Starzinski-Powitz, A., Rodt, H., Rollinghoff, M., Wagner, H.: Virus and TNP-hapten specific T cell mediated cytotoxicity against *H-2* incompatible target cells. J. Exp. Med. *143,* 999 (1976)

Rajewsky, K., Eichmann, K.: Antigen receptors of T helper cells. In: Contem. Top. Immunobiol. T Cells. *Stutman, O.* (Ed.). New York: Plenum Press, 1977, Volume 7, p. 69

Ramseier, H., Lindenmann, J.: F_1 hybrid animals: Reactivity against recognition structures of parental strain lymphoid cells. Pathol. Microbiol. *34,* 379 (1969)

Ramseier, H., Lindenmann, J.: Aliotypic antibodies. Transplant. Rev. *10,* 57 (1972)

Ramseier, H., Aguet, M., Lindenmann, J.: Similarity of idiotypic determinants of T- and B-lymphocyte receptors for alloantigens. Immunol. Rev. *34,* 50 (1977)

Rhen, T.G., Inman, J.K., Shearer, G.M.: Cell-mediated lympholysis to *H-2* matched target cells modified with a series of nitrophenyl compounds. J. Exp. Med. *144,* 1134 (1976)

Rosenthal, A.S., Shevach, E.M.: Function of macrophages in antigen recognition by guinea pig T lymphocytes. I. Requirement for histocompatible macrophages and lymphocytes. J. Exp. Med. *138,* 1194 (1973)

Rothenberg, B.E.: The self-recognition concept. Billups-Rothenberg, Inc. (1976)

Schmitt-Verhulst, A.-M., Shearer, G.M.: Bifunctional major histocompatibilitylinked genetic regulation of cell-mediated lympholysis to trinitrophenyl-modified autologous lymphocytes. J. Exp. Med. *142,* 914 (1975)

Schmitt-Verhulst, A.-M.,, Shearer, G.M.: Multiple *H-2*-linked immune response gene control of *H-2D*-associated T-cell-mediated lympholysis to trinitrophenyl-modified autologous cells: *Ir*-like genes mapping to the left of *I-A* and within the *I* region. J. Exp. Med. *144,* 1701 (1976)

Schrader, J.W., Edelman, G.M.: Joint recognition by cytotoxic T cells of inactivated Sendai virus and products of the major histocompatibility complex. J. Exp. Med. *145,* 523 (1977)

Schrader, J.W., Cunningham, B.A., Edelman, G.M.: Functional interactions of viral and histocompatibility antigens at tumor cell surfaces. Proc. Natl. Acad. Sci. USA *72,* 5066 (1975)

Schrader, J.W., Henning, R., Milner, R.J., Edelman, G.M.: The recognition of H-2 and viral antigens by cytotoxic T cells. In: Origins of Lymphocyte Diversity, Volume XLI, Cold Spring Harbor Symp. Quant. Biol. (1977)

Schwartz, R.N., Paul, W.E.: T-lymphocyte enriched murine peritoneal exudate cells. II. Genetic control of antigen induced T-lymphocyte proliferation. J. Exp. Med. *143,* 529 (1976)

Shearer, G.M.: Cell-mediated cytotoxicity to trinitrophenyl-modified syngeneic lymphocytes. Eur. J. Immunol. *4,* 257 (1974)

Shearer, G.M., Rehn, G.R., Garbarino, C.A.: Cell-mediated lympholysis of trinitrophenyl-modified autologous lymphocytes. Effector cell specificity to modified cell surface components controlled by the *H-2K* and *H-2D* regions of the murine major histocompatibility complex. J. Exp. Med. *141,* 1348 (1975)

Shearer, G.M., Rehn, T.G., Schmitt-Verhulst, A.-M.: Role of the murine major histocompatibility complex in the specificity of *in vitro* T-cell-mediated lympholysis against chemically-modified autologous lymphocytes. Transplant. Rev. *29,* 222–247 (1976)

Shearer, G.M., Schmitt-Verhulst, A.-M., Rehn, T.G.: Significance of the major histocompatibility complex as assessed by T-cell-mediated lympholysis involving syngeneic stimulating cells. In: Contemp. Top. Immunobiol. T Cells, *Stutman, O.* (Ed.). New York: Plenum Press 1977, Volume 7, p. 221

Shevach, E.M., Rosenthal, A.S.: Function of macrophages in antigen recognition by guinea pig T lymphocytes. II. Role of the macrophage in the regulation of genetic control of the immune response. J. Exp. Med. *138,* 1213 (1973)

Shreffler, D.C., David, C.S.: The *H-2* major histocompatibility complex and the *I* immune response region: genetic variation, function and organization. Adv. Immunol. *20,* 125 (1975)

Simonsen, M.: Graft versus host reactions. Their natural history and applicability as tools of research. Prog. Allergy *6,* 349–466 (1962)

Simonsen, M.: The lymphocytes in the regulation of immunological responses. The Heterogeneity of lymphocytes. Returns to square one. Allergology. Proc. of the 8th Congress of the Int. Assoc. of Allergology. Excerpta Med. Amsterdam (1974) p. 146

Simpson, E., Gordon, R.D.: Responsiveness to H-Y antigen, Ir gene complementation and target cell specificity. Transplant Rev., *35,* 59–75 (1977)

Singer, J.: personal communication (1977)

Snell, G.D.: The H-2 locus of the mouse: Observations and speculations concerning its comparative genetics and its polymorphism. Folia. Biol. *14*, 335 (1968)

Stulting, R.D., Todd, R.F., Gooding, L.R.: Susceptibility of anti-H-2-capped target cells to humoral and T lymphocyte-mediated lysis. Transplantation *21*, 71 (1976)

Svet-Moldavsky, G.J., Mkheidze, D.M., Liozner, A.L., Bykovsky, A.Ph.: Skin heterogenizing virus. Nature (London) *217*, 102 (1968)

Theodor, J.L.: Distinction between "self" and "not-self" in lower invertebrates. Nature (London) *227*, 690 (1970)

Thomas, L.: In: Cellular and Humoral Aspects of the Hypersensitive States. *Lawrence, H.S.* (Ed.) New York: Hoeber, 1959, p. 529

Thomas, D.W., Shevach, E.M.: Nature of the antigenic complex recognized by T lymphocytes: Specific sensitisation by antigens associated with allogeneic macrophages. Proc. Natl. Acad. Sci. USA *74*, 2104–2108 (1977)

Todaro, G.J.: Type C virogenes: genetic transfer and interspecies transfer. In: Tumor Virus Infections and Immunity, R.L. Crow, H. Friedman, J.E. Prier (Eds.). University Park Press, Baltimore, (1975), p. 35–44

Toivanen, P., Toivanen, A., Sorvari, T.: Incomplete restoration of the bursa-dependent immune system after transplantation of allogeneic stem cells into immunodeficient chicks. Proc. Natl. Acad. Sci. USA *71*, 957–961 (1974)

Vladutiu, A.O., Rose, N.R.: HLA antigens: Association with disease. Immunogenetics *1*, 305 (1972)

Wagner, H., Götze, D., Ptschelinzen, L., Röllinghoff, M.: Induction of cytotoxic T lymphocytes against *I*-region-coded determinants: *in vitro* evidence for a third histocompatibility locus in the mouse. J. Exp. Med. *142*, 1477 (1975)

Wainberg, M.A., Markson, Y., Weiss, D.W., Donjanski, F.: Cellular immunity against Rous sarcoma of chickens. Preferential reactivity against autochthonous target cells as determined by lymphocyte adherence and cytotoxicity tests *in vitro*. Proc. Natl. Acad. Sci. USA *71*, 3565 (1974)

Wilson, D.B., Lindahl, D.F., Wilson, D.H., Sprent, J.: The generation of killer cells to TNP-modified allogeneic targets by lymphocyte populations negatively selected to strong alloantigens. J. Exp. Med. *146*, 361–367 (1977)

Zinkernagel, R.M.: H-2 restriction of virus-specific cytotoxicity across the H-2 barrier. Separate effector T cell specificities are associated with self H-2 and with the tolerated allogeneic H-2 chimeras. J. Exp. Med. *144*, 933–945 (1976a)

Zinkernagel, R.M.: H-2 compatibility requirement for virus-specific T cell mediated cytolysis. The H-2K structure involved is coded by a single cistron defined by H-2K^b mutant mice. J. Exp. Med. *143*, 437–443 (1976b)

Zinkernagel, R.M.: Major transplantation antigens in T-cell mediated immunity: A comparison of the transplantation reaction with antiviral immunity. Fed. Proc., in press (1978)

Zinkernagel, R.M., Doherty, P.C.: Restriction of *in vitro* T cell mediated cytotoxicity in lymphocytic choriomeningitis within a syngeneic or semiallogeneic system. Nature (London) *248*, 701–702 (1974a)

Zinkernagel, R.M., Doherty, P.C.: Activity of sensitized thymus derived lymphocytes in lymphocytic choriomeningitis reflects immunological surveillance against altered self components. Nature (London) *251*, 547–548 (1974b)

Zinkernagel, R.M., Doherty, P.C.: H-2 compatibility requirement for T cell mediated lysis of targets infected with lymphocytic choriomeningitis virus. Different cytotoxic T cell specificities are associated with structures coded in H-2K or H-2D. J. Exp. Med. *141*, 1427–1436 (1975)

Zinkernagel, R.M., Doherty, P.C.: Possible mechanisms of disease susceptibility association with major transplantation antigens. In: HLA and Disease. *Dausset, J., Svejgaard, A.* (Eds.), Copenhagen: Munksgaard, 1976a

Zinkernagel, R.M., Doherty, P.C.: Does the apparent H-2 compatibility requirement for virus-specific T cell mediated cytolysis reflect T cell specificity for "altered self" or physiological interaction mechanisms? In: The Role of Products of the Histocompatibility Gene Complex in Immune Responses. *Katz, D.H., Benacerraf, B.* (Eds.). New York: Academic Press, 1976b, pp. 203–211

Zinkernagel, R.M., Doherty, P.C.: Major transplantation antigens virus and specificity of surveillance T cells. The "altered self" hypothesis. In: Contemp. Top. Immunobiol. *7*, 179–220 (1977a)

Zinkernagel, R.M., Doherty, P.C.: The concept that surveillance of self is mediated via the same set of genes that determines recognition of allogeneic cells. Cold Spring Harbor Lab. *XLI*, 505–510 (1977b)

Zinkernagel, R.M., Klein, J.: H-2-associated specificity of virus-immune cytotoxic T cells from H-2 mutant and wild-type mice: M523 (H-2K^{ka}) and M505 (H-2K^{bd}) do, M504 (H-2D^{da}) and M506 (H-2K^{fa}) do not crossreact with wild-type H-2K or H-2D. Immunogenetics, *4*, 581–590 (1977)

Zinkernagel, R.M., Welsh, R.M.: H-2 compatibility requirement for virus-spentic T-cell-mediated effector functions *in vivo*. I. Specificity of T cells conferring antiviral protection against lymphocytic choriomeningitis virus is associated with H-2K and H-2D. J. Immunol. *117*, 1495–1502 (1976)

Zinkernagel, R.M., Adler, B., Althage, A.: The question of derepression of H-2 specificities in virus-infected cells: Failure to detect specific alloreactive T cells after systemic virus infection or alloantigens detectable by alloreactive T cells on virus infected target cells. Immunogenetics, *5*, 361–378 (1977b)

Zinkernagel, R.M., Althage, A., Adler, B., Blanden, R.V., Davidson, W.F., Kees, U., Dunlop, M.B.C., Shreffler, D.C.: H-2 restriction of cell-mediated immunity to an intracellular bacterium. Effector T cells are specific for Listeria antigen in association with H-21 region coded self-markers. J. Exp. Med. *145*, 1353–1367 (1977c)

Zinkernagel, R.M., Althage, A., Jensen, F.C.: Cell-mediated immune response to lymphocytic choriomeningitis and vaccinia virus in rats. J. Immunol., *119*, 1242–1247 (1977a)

Zinkernagel, R.M., Callahan, G.N., Althage, A., Cooper, S., Klein, P.A., Klein, J.: On the thymus in the differentiation of "H-2 self-recognition" by T cells: Evidence for dual recognition? J. Exp. Med. *141*, 882–896 (1978a)

Zinkernagel, R.M., Callahan, G.N., Althage, A., Cooper, S., Streilein, J.W., Klein, J.: The lymphoreticular system in triggering virus-plus-self-specific cytotoxic T cells: Evidence for T help. J. Exp. Med. *147*, 897–911 (1978b)

Zinkernagel, R.M., Callahan, G.N., Klein, J., Dennert, G.: Cytotoxic T cells learn specificity for self H-2 during differentiation in the thymus. Nature (London) *271*, 251–253 (1978c)

Other Volumes of Interest from this Series

Volume 81
Lymphocyte Hybridomas

Editors: F. Melchers, M. Potter, N. Warner

Second Workshop on "Functional Properties of Tumors of T and B Lymphocytes"

Sponsored by the National Cancer Institute April 3–5, 1978 Bethesda, MD, USA

1978. 85 figures, 66 tables, 255 pages
ISBN 3-540-08810-5

Volume 80

1978. 31 figures, 16 tables, IV, 170 pages
ISBN 3-540-08781-8

Contents: J.J. Bullen, H.J. Rogers, E. Griffiths: Role of Iron in Bacterial Infection. — J.H.L. Playfair: Effective and Ineffective Immune Responses to Parasites: Evidence from Experimental Models. — G.R. Pearson: In Vitro and in Vivo Investigations on Antibody-Dependent Cellular Cytotoxicity. — S. Cohen, G.H. Mitchell: Prospects for Immunisation Against Malaria. — C. Scholtissek: The Genome of the Influenza Virus.

Volume 79

35 figures, 29 tables, III, 309 pages. 1978
ISBN 3-540-08587-4

Contents: H. Fan: Expression of RNA Tumor Viruses at Translation and Transcription Levels. — M. Mussgay, O.-R. Kaaden: Progress in Studies on the Etiology and Serologic Diagnosis of Enzootic Bovine Leukosis. — K.L. Beemon: Oligonucleotide Fingerprinting With RNA Tumors Virus RNA. — J.A. Levy: Xenotropic Type C-Viruses. — M.B. Gardner: Type C Viruses of Wild Mice: Characterization and Natural History of Amphotropic, Ecotropic, and Xenotropic MuLV. — R.R. Friis: Temperature-Sensitive Mutants of Avian RNA Tumor Viruses: A Review. — E. Hunter: The Mechanism for Genetic Recombination in the Avian Retroviruses.

Volume 78

45 figures, 11 tables, III, 248 pages. 1977
ISBN 3-540-08499-1

Contents: H. zur Hausen: Human Papillomaviruses and Their Possible Role in Squamous Cell Carcinomas. — G.G.B. Klaus, A.K. Abbas: Antigen-Receptor Interaction in the Induction of B-Lymphocyte Unresponsiveness. — T. Hohn, I. Katsura: Structure and Assembly of Bacteriophage Lambda. — S.A. Plotkin: Perinatally Acquired Viral Infections. — J. Collins: Gene Cloning with Small Plasmids. — H.A. Nash: Integration and Excision of Bacteriophage λ. — A.M. Skalka: DNA Replication — Bacteriophage Lambda. — G. Wengler: Structure and Function of the Genome of Viruses Containing Single-Stranded RNA as Genetic Material: The Concept of Transcription and Translation Helices and the Classification of these Viruses into Six Groups.

Volume 77

19 figures, III, 168 pages. 1977
ISBN 3-540-08401-0

Contents: B.E. Butterworth: Proteolytic Processing of Animal Virus Proteins. — K. Kano, F. Milgrom: Heterophile Antigens and Antibodies in Medicine. — W.E. Rawls, S. Bacchetti, F.L. Graham: Relation of Herpes Simplex Viruses to Human Malignancies. — W. Hengstenberg: Enzymology of Carbohydrate Transport in Bacteria. — A.E. Butterworth: The Eosinophil and its Role in Immunity to Helminth Infection.

Reviews of Physiology, Biochemistry and Pharmacology

formerly
Ergebnisse der Physiologie, biologischen Chemie und experimentellen Pharmakologie
Editors: R.H. Adrian, E. Helmreich, H. Holzer, R. Jung, O. Krayer, R.J. Linden, F. Lynen, P.A. Miescher, J. Piiper, H. Rasmussen, A.E. Renold, U. Trendelenburg, K. Ullrich, W. Vogt, A. Weber

This series presents rapid and comprehensive information on topical problems and research in progress over the entire range of physiology, biochemistry, and pharmacology. An international group of editors is responsible for inviting experts in these fields to submit contributions. Every year, three to four volumes are published. The language of publication is English.

Volume 83

1978. 45 figures, 15 tables. Approx. 200 pages
ISBN 3-540-08907-1

Contents: E.M. Wright: Transport Processes in the Formation of the Cerebrospinal Fluid. — L.B. Cohen, B.M. Salzberg: Optical Measurement of Membrane Potential. — L. Glaser: Cell-Cell Adhesion Studies with Embryonal and Cultured Cells. — P. Propping: Pharmacogenetics. — Author Index. — Subject Index.

Volume 82

43 figures, 17 tables, V, 214 pages
ISBN 3-540-08748-6

Contents: A. Levitzki: Catecholamine Receptors. — G. Schreiber, J. Urban: The Synthesis and Secretion of Albumin. — W. Almers: Gating Currents and Charge Movements in Excitable Membranes. — Author Index. — Subject Index.

Volume 81

41 figures, 8 tables, V, 214 pages. 1978
ISBN 3-540-08554-8

Contents: R.E. Langmann: Cell-Mediated Immunity and the Major Histocompatibility Complex. — W. Kobinger: Central α-Adrenergic Systems as Targets for Hypotensive Drugs. — P.C. Heinrich, V. Gross, W. Northemann, M. Scheurlen: Structure and Function of Nuclear Ribonucleoprotein Complexes. — L. Hösli, E. Hösli: Action and Uptake of Neurotransmitters in CNS Tissue Culture. — Author Index. — Subject Index. — Indexed in Current Contents.

Volume 80

1977. 44 figures, 2 tables, V, 197 pages
ISBN 3-540-08466-5

Contents: C. Baylis, B.M. Brenner: The Physiologic Determinants of Glomerular Ultrafiltration. — H. Meinhardt: Models for the Ontogenetic Development of Higher Organisms. — W. Burke, A.M. Cole: Extraretinal Influences on the Lateral Geniculate Nucleus. — Author Index. — Subject Index. — Author and Subject Index of Volumes 50–80. — Indexed in Current Contents.

SPRINGER-VERLAG
BERLIN HEIDELBERG NEW YORK